Ultrasound B-Mode Imaging

Ultrasound B-Mode Imaging: Beamforming and Image Formation Techniques

Special Issue Editors

Giulia Matrone
Alessandro Ramalli
Piero Tortoli

MDPI • Basel • Beijing • Wuhan • Barcelona • Belgrade

Special Issue Editors

Giulia Matrone
University of Pavia
Italy

Alessandro Ramalli
KU Leuven
Belgium

Piero Tortoli
University of Florence
Italy

Editorial Office
MDPI
St. Alban-Anlage 66
4052 Basel, Switzerland

This is a reprint of articles from the Special Issue published online in the open access journal *Applied Sciences* (ISSN 2076-3417) from 2018 to 2019 (available at: https://www.mdpi.com/journal/applsci/special_issues/Ultrasound_B-mode_Imaging).

For citation purposes, cite each article independently as indicated on the article page online and as indicated below:

LastName, A.A.; LastName, B.B.; LastName, C.C. Article Title. *Journal Name* **Year**, *Article Number*, Page Range.

ISBN 978-3-03921-199-9 (Pbk)
ISBN 978-3-03921-200-2 (PDF)

© 2019 by the authors. Articles in this book are Open Access and distributed under the Creative Commons Attribution (CC BY) license, which allows users to download, copy and build upon published articles, as long as the author and publisher are properly credited, which ensures maximum dissemination and a wider impact of our publications.

The book as a whole is distributed by MDPI under the terms and conditions of the Creative Commons license CC BY-NC-ND.

Contents

About the Special Issue Editors vii

Giulia Matrone, Alessandro Ramalli and Piero Tortoli
Ultrasound B-Mode Imaging: Beamforming and Image Formation Techniques
Reprinted from: *Appl. Sci.* **2019**, *9*, 2507, doi:10.3390/app9122507 1

Libertario Demi
Practical Guide to Ultrasound Beam Forming:
Beam Pattern and Image Reconstruction Analysis
Reprinted from: *Appl. Sci.* **2018**, *8*, 1544, doi:10.3390/app8091544 5

Sua Bae and Tai-Kyong Song
Methods for Grating Lobe Suppression in Ultrasound Plane Wave Imaging
Reprinted from: *Appl. Sci.* **2018**, *8*, 1881, doi:10.3390/app8101881 20

Ling Tong, Qiong He, Alejandra Ortega, Alessandro Ramalli, Piero Tortoli, Jianwen Luo and Jan D'hooge
Coded Excitation for Crosstalk Suppression in Multi-line Transmit Beamforming: Simulation Study and Experimental Validation
Reprinted from: *Appl. Sci.* **2019**, *9*, 486, doi:10.3390/app9030486 28

Giulia Matrone and Alessandro Ramalli
Spatial Coherence of Backscattered Signals in Multi-Line Transmit Ultrasound Imaging and Its Effect on Short-Lag Filtered-Delay Multiply and Sum Beamforming
Reprinted from: *Appl. Sci.* **2018**, *8*, 486, doi:10.3390/app8040486 44

Maxime Polichetti, François Varray, Jean-Christophe Béra, Christian Cachard and Barbara Nicolas
A Nonlinear Beamformer Based on p-th Root Compression—Application to Plane Wave Ultrasound Imaging
Reprinted from: *Appl. Sci.* **2018**, *8*, 599, doi:10.3390/app8040599 59

Ken Inagaki, Shimpei Arai, Kengo Namekawa and Iwaki Akiyama
Sound Velocity Estimation and Beamform Correction by Simultaneous Multimodality Imaging with Ultrasound and Magnetic Resonance
Reprinted from: *Appl. Sci.* **2018**, *8*, 2133, doi:10.3390/app8112133 74

Chang Liu, Binzhen Zhang, Chenyang Xue, Wendong Zhang, Guojun Zhang and Yijun Cheng
Multi-Perspective Ultrasound Imaging Technology of the Breast with Cylindrical Motion of Linear Arrays
Reprinted from: *Appl. Sci.* **2019**, *9*, 419, doi:10.3390/app9030419 85

Mohamed Yaseen Jabarulla and Heung-No Lee
Speckle Reduction on Ultrasound Liver Images Based on a Sparse Representation over a Learned Dictionary
Reprinted from: *Appl. Sci.* **2018**, *8*, 903, doi:10.3390/app8060903 96

Wei Guo, Yusheng Tong, Yurong Huang, Yuanyuan Wang and Jinhua Yu
A High-Efficiency Super-Resolution Reconstruction Method for Ultrasound Microvascular Imaging
Reprinted from: *Appl. Sci.* **2018**, *8*, 1143, doi:10.3390/app8071143 113

Monika Makūnaitė, Rytis Jurkonis, Alberto Rodríguez-Martínez, Rūta Jurgaitienė, Vytenis Semaška, Karolina Mėlinytė and Raimondas Kubilius
Ultrasonic Parametrization of Arterial Wall Movements in Low- and High-Risk CVD Subjects
Reprinted from: *Appl. Sci.* **2019**, *9*, 465, doi:10.3390/app9030465 . . . . **125**

About the Special Issue Editors

Giulia Matrone received her B.Sc. and M.Sc. degrees in Biomedical Engineering, both cum laude, from the University of Pavia, Pavia, Italy, in 2006 and 2008 respectively, and a Ph.D. degree in Bioengineering and Bioinformatics from the same university in 2012. From 2012 to 2016, she was a Postdoctoral Researcher with the Bioengineering Laboratory, Department of Electrical, Computer and Biomedical Engineering, University of Pavia, where she is currently Assistant Professor of Bioengineering. Her research interests are mainly in the field of ultrasound medical imaging and signal processing, and include beamforming and image formation techniques, simulations, system-level analyses for the design of 3D ultrasound imaging probes, ultrasound elastography, and microwave imaging for biomedical applications.

Alessandro Ramalli was born in Prato, Italy, in 1983. He received his Master's degree in electronics engineering from the University of Florence, Florence, Italy, in 2008. His Ph.D. degree was awarded in 2012 as the result of a joint project conducted in electronics system engineering at the University of Florence, and in automation, systems, and images at the University of Lyon. From 2012 to 2017, he was involved in the development of the imaging section of a programmable open ultrasound system. He is currently a Postdoctoral Researcher with the Laboratory of Cardiovascular Imaging and Dynamics, KU Leuven, Leuven, Belgium, granted by the European Commission through a "Marie Skłodowska-Curie Individual Fellowship". His current research interests include medical imaging, beamforming methods, and ultrasound simulation.

Piero Tortoli received the Laurea degree in electronics engineering from the University of Florence, Italy, in 1978. Since then, he has been on the faculty of the Electronics and Telecommunications (now Information Engineering) Department of the University of Florence, where he is currently full Professor of Electronics, leading a group of about 10 researchers in the Microelectronics Systems Design Laboratory. His research interests include the development of open ultrasound research systems and novel imaging/Doppler methods. He has authored more than 260 papers on these topics. Professor Tortoli has served on the IEEE International Ultrasonics Symposium Technical Program Committee since 1999, and is currently Associate Editor of the IEEE Transactions on Ultrasonics, Ferroelectrics, and Frequency Control. He chaired the 22nd International Symposium on Acoustical Imaging (1995), the 12th New England Doppler Conference (2003), established the Artimino Conference on Medical Ultrasound in 2011, and organized it again in 2017. In 2000, he was named an Honorary Member of the Polish Academy of Sciences. He has been an elected Member of the Academic Senate at the University of Florence since 2016.

Editorial

Ultrasound B-Mode Imaging: Beamforming and Image Formation Techniques

Giulia Matrone [1],*, Alessandro Ramalli [2,3,*] and Piero Tortoli [3,*]

1. Department of Electrical, Computer and Biomedical Engineering, University of Pavia, 27100 Pavia, Italy
2. Department of Cardiovascular Imaging and Dynamics, KU Leuven, 3000 Leuven, Belgium
3. Department of Information Engineering, University of Florence, 50139 Florence, Italy
* Correspondence: giulia.matrone@unipv.it (G.M.); alessandro.ramalli@kuleuven.be (A.R.); piero.tortoli@unifi.it (P.T.)

Received: 7 June 2019; Accepted: 16 June 2019; Published: 19 June 2019

1. Introduction

In the last decade, very active research in the field of ultrasound medical imaging has brought to the development of new advanced image formation techniques and of high-performance systems able to effectively implement them [1]. For years, Brightness (B)-mode, one of the mostly used ultrasound imaging modalities [2], has been based on a time-consuming process, in which focused beams are iteratively sent into the body and the received waves are used to form an image scan-line, covering line-by-line the region of interest.

"Image formation" refers to the whole process of image reconstruction, starting from the transmission strategy to the reception of signals, beamforming, and image processing. The role of the so-called "beamformer" is central in this process, as it manages the ultrasound beam generation, focusing, and steering [3]. Image quality is in fact deeply influenced by the beam shape, and thus the beamforming optimization plays an important role in maximizing the signal-to-noise ratio, contrast, and resolution of the final image, while limiting as much as possible off-axis interferences to reject clutter and noise. Additionally, an important goal is to improve the acquisition frame-rate, which, as mentioned above, is limited by the line-by-line acquisition process [4].

Image enhancement methods play an important role during both the image pre- and post-processing phases [5]. In the former case, these techniques aim at improving the quality of B-mode frames by directly operating on the image generation process, as for example in spatial/frequency compounding, pulse compression, or harmonic imaging. The latter category instead refers to approaches aimed at reducing noise/artifacts, making speckle more uniform, detecting edges, and consequently facilitating the following processing steps, like segmentation or measurement of quantitative parameters.

Given the above premises, this Special Issue was launched to collect novel contributions on both ultrasound beamforming and image formation techniques. Twenty-one interesting works were consequently submitted and, among them, 10 were selected for publication (i.e., 48% acceptance rate).

2. Ultrasound B-Mode Imaging

The Special Issue opens with a review paper on the main ultrasound beamforming techniques [6]. The classic beamforming method for linear/phase array imaging is first introduced, before presenting advanced methods: from multi-line transmission and acquisition to synthetic aperture imaging, passing through plane wave, and diverging wave imaging. The stress is on the peculiarity of each method in terms of spatio-temporal resolution, contrast, penetration depth, aperture size, and field of view. The paper may represent a useful handbook for users who need to choose the most appropriate beamforming method for the specific application of interest.

The following nine papers are grouped in three main groups, dealing with novel beamforming techniques, non-conventional image formation, and image enhancement, respectively.

2.1. Novel Beamforming Techniques

High frame-rate imaging techniques [4] have recently gained increased interest for their capability to detect fast dynamic events. However, the improvement of temporal resolution comes at the expense of image quality, thus pushing researchers to recover it by developing smart strategies. Four papers have been published in this Special Issue presenting advanced transmission sequences [7,8] and beamforming schemes [9,10] applied to either plane waves [7,10] or multi-line transmission imaging [8,9].

Bae and Song [7] analyzed the grating lobe artifacts due to the compounding of images obtained from the transmission of steered plane waves with a constant angle interval. Additionally, they showed that the use of non-uniform angle sets is a smart solution to keep the frame rate high, while limiting the level of image artifacts due to grating lobes. Tong et al. [8] studied the effectiveness of orthogonal coded excitations in multi-line transmission imaging in suppressing crosstalk artifacts. They showed that Golay codes enable higher crosstalk rejection (and better contrast) compared to linear chirps.

Two papers focus on the so-called coherence-based beamforming methods. Spatial coherence of ultrasound backscattered echoes is affected by contributions coming from off-axis regions, noise, and interferences. Matrone and Ramalli (Guest Editors) presented a new formulation of the Filtered Delay Multiply and Sum (F-DMAS) beamforming, namely Short-Lag F-DMAS [9]. They provided new insights into the relation between the performance of the F-DMAS algorithm and the coherence of backscattered signals in multi-line transmission imaging. Polichetti et al. presented a generalized and extended formulation of the F-DMAS beamformer, referred to as p-DAS [10]. They applied the proposed method to plane wave imaging and showed the achieved improvements in terms of lateral resolution and artifacts rejection.

2.2. Non-Conventional Image Formation

Non-conventional imaging systems have been proposed to improve the B-mode image quality and its diagnostic content. As an example, Inagaki et al. [11] designed and built a multi-modality (ultrasound and magnetic resonance) system to estimate the ultrasound propagation speed in the region of interest. The estimates were then used to correct the beamforming delay, both in transmission and in reception, thus enhancing the image resolution and signal-to-noise ratio. Liu et al. [12] proposed a multi-perspective ultrasound imaging system based on four 3.5 MHz linear arrays. These arrays were placed, in a cross shape, on a motorized rotatory table to perform 3D ultrasound computed tomography of a breast model with different inclusions. The boundary of the breast, as well as the inclusions, could be clearly seen from all the perspectives, hence potentially improving the specificity and sensitivity of ultrasonic diagnosis.

2.3. Image Enhancement

Image quality enhancement can also be obtained through post-processing methods for image filtering, deconvolution, tracking, segmentation, and tissue characterization. In this Special Issue, Jabarulla and Lee [13] proposed a technique for liver images based on a signal reconstruction model, known as sparse representation over dictionary learning. This technique allows filtering the speckle while preserving the image features and the edges of anatomical structures. Guo et al. [14] presented a novel super-resolution reconstruction method. They developed a low computational load technique for microbubble localization and trajectory tracking. They showed that the proposed method improves the image resolution by using fewer frames than other reference methods, thus moving super-resolution a step forward to real-time imaging. Makūnaitė et al. [15] showed how advanced segmentation and tracking techniques can be exploited to develop new predictors of cardiovascular events. Specifically, they tracked arterial wall movements for the evaluation of arterial stiffness and showed that the average value of the intima-media thickness, during the cardiac cycle, is statistically different between healthy volunteers and patients at risk of cardiovascular disease.

3. Future Perspectives

The different contributions published in this Special Issue confirm that the research of new strategies to improve the image formation process keeps on being a hot topic in the ultrasound imaging community. In this sense, it is also worth pointing out that efforts have been recently devoted to objectively evaluating and comparing novel beamforming methods, by creating development/test platforms and datasets [16,17] to be shared by all research groups working on ultrasound beamforming.

Further active research is thus expected in this field, where many challenges still persist, especially when dealing with the difficult-to-image patients. For this reason, efforts should always be supported by real clinical needs, and image enhancement should be aimed at increasing visibility of anatomical structures and easing image interpretation and clinical parameters extraction, towards a more and more effective diagnostic process. An increasing involvement of clinicians in the in vivo evaluation of real image quality from a medical point of view is thus desirable.

Acknowledgments: The Guest Editors wish to thank all the authors who have submitted papers to this Special Issue and all the reviewers who allowed improving the quality of the submitted manuscripts by working with dedication and timeliness. Finally, we gratefully thank the editorial team of Applied Sciences and Daria Shi, our Assistant Editor, for their extraordinary support.

Conflicts of Interest: The authors declare no conflict of interest.

References

1. Boni, E.; Yu, A.C.H.; Freear, S.; Jensen, J.A.; Tortoli, P. Ultrasound Open Platforms for Next-Generation Imaging Technique Development. *IEEE Trans. Ultrason. Ferroelectr. Freq. Control* **2018**, *65*, 1078–1092. [CrossRef] [PubMed]
2. Szabo, T.L. *Diagnostic Ultrasound Imaging: Inside Out*, 1st ed.; Academic Press: Cambridge, MA, USA, 2004; ISBN 0-12-680145-2.
3. Van Veen, B.; Buckley, K.M. *Wireless, Networking, Radar, Sensor Array Processing, and Nonlinear Signal Processing*, 1st ed.; CRC Press: Boca Raton, FL, USA, November 2009; Volume Beamforming techniques for spatial filtering.
4. Tanter, M.; Fink, M. Ultrafast imaging in biomedical ultrasound. *IEEE Trans. Ultrason. Ferroelectr. Freq. Control* **2014**, *61*, 102–119. [CrossRef] [PubMed]
5. Contreras Ortiz, S.H.; Chiu, T.; Fox, M.D. Ultrasound image enhancement: A review. *Biomed. Signal Process. Control* **2012**, *7*, 419–428. [CrossRef]
6. Demi, L. Practical Guide to Ultrasound Beam Forming: Beam Pattern and Image Reconstruction Analysis. *Appl. Sci.* **2018**, *8*, 1544. [CrossRef]
7. Bae, S.; Song, T.-K. Methods for Grating Lobe Suppression in Ultrasound Plane Wave Imaging. *Appl. Sci.* **2018**, *8*, 1881. [CrossRef]
8. Tong, L.; He, Q.; Ortega, A.; Ramalli, A.; Tortoli, P.; Luo, J.; D'hooge, J. Coded Excitation for Crosstalk Suppression in Multi-Line Transmit Beamforming: Simulation Study and Experimental Validation. *Appl. Sci.* **2019**, *9*, 486. [CrossRef]
9. Matrone, G.; Ramalli, A. Spatial Coherence of Backscattered Signals in Multi-Line Transmit Ultrasound Imaging and Its Effect on Short-Lag Filtered-Delay Multiply and Sum Beamforming. *Appl. Sci.* **2018**, *8*, 486. [CrossRef]
10. Polichetti, M.; Varray, F.; Béra, J.-C.; Cachard, C.; Nicolas, B. A Nonlinear Beamformer Based on p-th Root Compression—Application to Plane Wave Ultrasound Imaging. *Appl. Sci.* **2018**, *8*, 599. [CrossRef]
11. Inagaki, K.; Arai, S.; Namekawa, K.; Akiyama, I. Sound Velocity Estimation and Beamform Correction by Simultaneous Multimodality Imaging with Ultrasound and Magnetic Resonance. *Appl. Sci.* **2018**, *8*, 2133. [CrossRef]
12. Liu, C.; Zhang, B.; Xue, C.; Zhang, W.; Zhang, G.; Cheng, Y. Multi-Perspective Ultrasound Imaging Technology of the Breast with Cylindrical Motion of Linear Arrays. *Appl. Sci.* **2019**, *9*, 419. [CrossRef]

13. Jabarulla, M.Y.; Lee, H.-N. Speckle Reduction on Ultrasound Liver Images Based on a Sparse Representation over a Learned Dictionary. *Appl. Sci.* **2018**, *8*, 903. [CrossRef]
14. Guo, W.; Tong, Y.; Huang, Y.; Wang, Y.; Yu, J. A High-Efficiency Super-Resolution Reconstruction Method for Ultrasound Microvascular Imaging. *Appl. Sci.* **2018**, *8*, 1143. [CrossRef]
15. Makūnaitė, M.; Jurkonis, R.; Rodríguez-Martínez, A.; Jurgaitienė, R.; Semaška, V.; Mėlinytė, K.; Kubilius, R. Ultrasonic Parametrization of Arterial Wall Movements in Low-and High-Risk CVD Subjects. *Appl. Sci.* **2019**, *9*, 465. [CrossRef]
16. Liebgott, H.; Rodriguez-Molares, A.; Cervenansky, F.; Jensen, J.A.; Bernard, O. Plane-Wave Imaging Challenge in Medical Ultrasound. In Proceedings of the 2016 IEEE International Ultrasonics Symposium (IUS), Tours, France, 18–21 September 2016; pp. 1–4.
17. Rodriguez-Molares, A.; Rindal, O.M.H.; Bernard, O.; Nair, A.; Bell, M.A.L.; Liebgott, H.; Austeng, A.; Løvstakken, L. The UltraSound ToolBox. In Proceedings of the 2017 IEEE International Ultrasonics Symposium (IUS), Washington, DC, USA, 6–9 September 2017; pp. 1–4.

© 2019 by the authors. Licensee MDPI, Basel, Switzerland. This article is an open access article distributed under the terms and conditions of the Creative Commons Attribution (CC BY) license (http://creativecommons.org/licenses/by/4.0/).

Review

Practical Guide to Ultrasound Beam Forming: Beam Pattern and Image Reconstruction Analysis

Libertario Demi

Department of Information Engineering and Computer Science, University of Trento, 38123 Trento, Italy; libertario.demi@unitn.it

Received: 9 August 2018; Accepted: 1 September 2018; Published: 3 September 2018

Abstract: Starting from key ultrasound imaging features such as spatial and temporal resolution, contrast, penetration depth, array aperture, and field-of-view (FOV) size, the reader will be guided through the pros and cons of the main ultrasound beam-forming techniques. The technicalities and the rationality behind the different driving schemes and reconstruction modalities will be reviewed, highlighting the requirements for their implementation and their suitability for specific applications. Techniques such as multi-line acquisition (MLA), multi-line transmission (MLT), plane and diverging wave imaging, and synthetic aperture will be discussed, as well as more recent beam-forming modalities.

Keywords: medical ultrasound; beam forming; ultrasound imaging; multi-line acquisition; multi-line transmission; plane wave; diverging wave; synthetic aperture; parallel beam forming; beam pattern; image reconstruction

1. Introduction

In ultrasound medical imaging, beam forming in essence deals with the shaping of the spatial distribution of the pressure field amplitude in the volume of interest, and the consequent recombination of the received ultrasound signals for the purpose of generating images. One can thus navigate through the different techniques using the following question as a compass: which imaging features are important to my application of interest, and which features can I sacrifice? There is, in fact, no ultimate beam-forming approach, and the answer to the previous question strongly depends on what one wants to see in the images.

Below are the key imaging features that will be considered in this paper to review the different beam-forming techniques, along with their descriptions:

Spatial resolution: the smallest spatial distance for which two scatterers can be distinguished in the final image. Spatial resolution can be either axial (along the direction of propagation of the ultrasound wave), lateral, or elevation resolution (along the plane to which the direction of propagation is perpendicular). This feature is normally expressed in mm.

Temporal resolution: the time interval between two consecutive images. This feature is normally expressed in Hz.

Contrast: the capability to visually delineate different objects, e.g., different tissue types, in the generated images. This feature is generally expressed in dB, and it is a relative measure between image intensities.

Penetration depth: the larger depths for which a sufficiently high signal-to-noise ratio (SNR) level can be maintained. This feature is normally expressed in cm.

Array aperture: the physical sizes of the surface representing the combined distribution of active and passive ultrasound sensors: in other words, the array footprint. The array aperture is defined by the number of ultrasound sensors (elements), their sizes, and their distribution. This feature is generally expressed in cm^2.

Field of view (FOV): the sizes of the area represented by the obtained images. This feature is generally expressed in cm² or cm³.

Although introduced individually, these features are strongly related. For example, a decrease in temporal resolution can be traded to achieve a higher spatial resolution or a larger FOV; a deeper penetration depth can be achieved by lowering the transmitted center frequency, thus deteriorating spatial resolution, or a broader insonification area can be achieved by widening the transmitted beam, which will result in lower pressure levels being generated, thus lowering the SNR compared to a focused beam. To help the reader become more familiar with these concepts, a simple model can be used. Assuming linear propagation, the following wave equation can be applied to model the pressure field generated by an arbitrary source which propagates in a homogeneous medium [1]:

$$\partial_x^2 p(x,t) - \frac{1}{c_0^2} \partial_t^2 p(x,t) = S(x,t) \qquad (1)$$

Here, ∂_x^2 and ∂_t^2 represent the second-order derivative as regards space and time, respectively, $p(x,t)$ is the pressure field, t is time, $x = (x,y,z)$ is the three-dimensional spatial coordinate in a Cartesian system, c_0 is the small signal speed of sound, and $S(x,t)$ is the source. For a monochromatic point source, i.e., $S(x,t) = \delta(x)\cos(2\pi f_0 t)$, the solution to this equation is known [1], and can be expressed as:

$$p_{MPs}(x,t) = \frac{P_0}{4\pi |x|} \cos\left[2\pi f_0 \left(t - \frac{|x|}{c_0}\right)\right] \qquad (2)$$

In this equation, $p_{MPs}(x,t)$ is the pressure field generated by the monochromatic point source, P_0 is the source amplitude, and f_0 is the source frequency. This solution is useful, because the pressure field generated by every source can be approximated as the sum of the pressure generated by several point sources, the position of which models the actual shape of the source. The following equation can then be applied:

$$p(x,t) = \sum_{i=1}^{N} \frac{P_0}{4\pi D_i} \cos\left[2\pi f_0 \left(t - \frac{D_i}{c_0}\right)\right], \qquad (3)$$

with N being the number of point sources, D_i being the distance between the point for which the pressure field is calculated, and the source being i. Equation (3) can also be expressed in its complex formulation as follows:

$$p(x,t) = \sum_{i=1}^{N} \frac{P_0}{4\pi D_i} e^{-j2\pi f_0 (t - \frac{D_i}{c_0})} = P_0 e^{-j2\pi f_0 t} \sum_{i=1}^{N} \frac{e^{j2\pi f_0 \frac{D_i}{c_0}}}{4\pi D_i} \propto \sum_{i=1}^{N} \frac{e^{j2\pi f_0 \frac{D_i}{c_0}}}{4\pi D_i}. \qquad (4)$$

The maximum pressure at a given location is thus obtained when the distances D_i are all the same, i.e., the point sources are placed on the surface of a sphere with radius r and centered at the location where the pressure field is calculated. Alternatively, in a case where the sources cannot be arranged in that way, each source could be multiplied by a phase coefficient that compensates for the differences between each term D_i. In essence, this means time delaying the source according to its distance from the point where the pressure field is calculated. Moreover, to increase the pressure field amplitude, one can increase the physical size of the actual source (the aperture), which entails increasing the number of point sources that are needed to describe it. From Equation (4), we can thus conclude that by applying appropriate phase coefficients, we can maximize the pressure field generated by an arbitrarily shaped source, and that the larger the aperture, the higher the pressure field. Generating high amplitudes means improving the signal strength, and thus the SNR.

If we then associate a specific phase and amplitude to each source, and model this as a function of the spatial coordinates $A(x)$, we can reformulate Equation (4) as:

$$p(x,t) = e^{-j2\pi f_0 t} \sum_{i=1}^{N} \frac{A(x)}{4\pi D_i} e^{j2\pi f_0 \frac{D_i}{c_0}}. \tag{5}$$

If we then assume that the source lies on the plane $z = 0$, and that the point with coordinates $x = (x, y, z)$ lies on a plane parallel to the plane $z = 0$, and at distance L from it, with L being much larger than the maximum distance between two point sources inside the planar surface representing the actual source, then we can write:

$$D_i = \sqrt{L^2 + (xi - X)^2 + (y_i - Y)^2} \tag{6}$$

with (x_i, y_i) being the coordinates of each point source, and (X, Y) being the coordinates describing the point x on the plane parallel to the plane $z = 0$. Using a binomial expansion, Equation (6) can be rewritten as:

$$D_i = L\left(1 + \frac{x_i^2 + X^2 - 2x_iX + y_i^2 + Y^2 - 2y_iY}{2L^2}\right) \tag{7}$$

and assuming $L \gg X, Y \gg x_i, y_i$ we can approximate:

$$D_i = L\left(1 + \frac{X^2 - 2x_iX + Y^2 - 2y_iY}{2L^2}\right) \tag{8}$$

Combining Equation (8) with Equation (5) we obtain:

$$p(X, Y, t) = \frac{e^{-j2\pi f_0 t} e^{\frac{j2\pi f_0}{c_0} L(1 + \frac{X^2 + Y^2}{2L^2})}}{4\pi L} \sum_x \sum_y A(x, y) e^{-\frac{j2\pi(xX + yY)}{\lambda_0 2L}}. \tag{9}$$

with $A(x, y) = 0$ where there is no source. Thus, the pressure field is proportional to the two-dimensional discrete Fourier transform of the function describing the source. Note that we have approximated D_i with L as regards the amplitude term inside the summation in Equation (5). This was not the case for the phase term. In fact, in this case also, small variations of D_i with respect to $\lambda_0 = \frac{c_0}{f_0}$ can be significant. From Equation (9), we can deduce that for a circular aperture with radius R, we can write the pressure field as:

$$p(X, Y, t)|_{Y=0} \propto \text{sinc}\left(\frac{RXf_0}{c_0 2L}\right) \tag{10}$$

From Equation (10), we can deduce that the ultrasound beam size is influenced by the aperture size and transmitted frequency, and that it changes over depth. The beam can be defined as the area where the pressure amplitude is above a specific value, which is normally considered in relation to the maximum pressure generated (e.g., the -20 dB beam). A larger aperture and higher frequencies mean a smaller beam. Moreover, the beam generally widens with increasing depths. The beam size defines the spatial resolution in the lateral and elevation direction. The smaller the beam, the higher the spatial resolution. On the other hand, the smaller the beam, the smaller the volume that can be insonified with a single transmission, and more transmission events are thus required to cover a given volume. In the next section, the basic differences between linear and phased array beam forming will be introduced. Subsequently, multi-line acquisition (MLA), multi-line transmission (MLT), plane and diverging wave imaging, synthetic aperture, and more recent beam-forming modalities will be described. To summarize the analysis, a table is presented where the peculiarities of each modality are highlighted.

2. Linear and Phased Array Beam Forming

We can start by describing the source that generates the ultrasound fields. In particular, we will address its aperture and how we could excite it by means of electrical signals. As represented in Figure 1, ultrasound sensors, which are generally able to both transmit and receive ultrasound signals, can be arranged so that their centers cover a surface, a line, or a curve. In the first case, we have a matrix or two-dimensional (2D) array, while in the second and third case, we have a one-dimensional (1D) array. The distance between the centers is referred to as pitch, and the size of the empty space between consecutive sensors is called kerf [2,3]. For 2D arrays, the pitch and kerf may be different along the lateral (x) and elevation (y) directions. In general, sensors do not need to be arranged in a periodic structure. In fact, an aperiodic sensor distribution can produce benefits such as the reduction of the effects of side lobes [4].

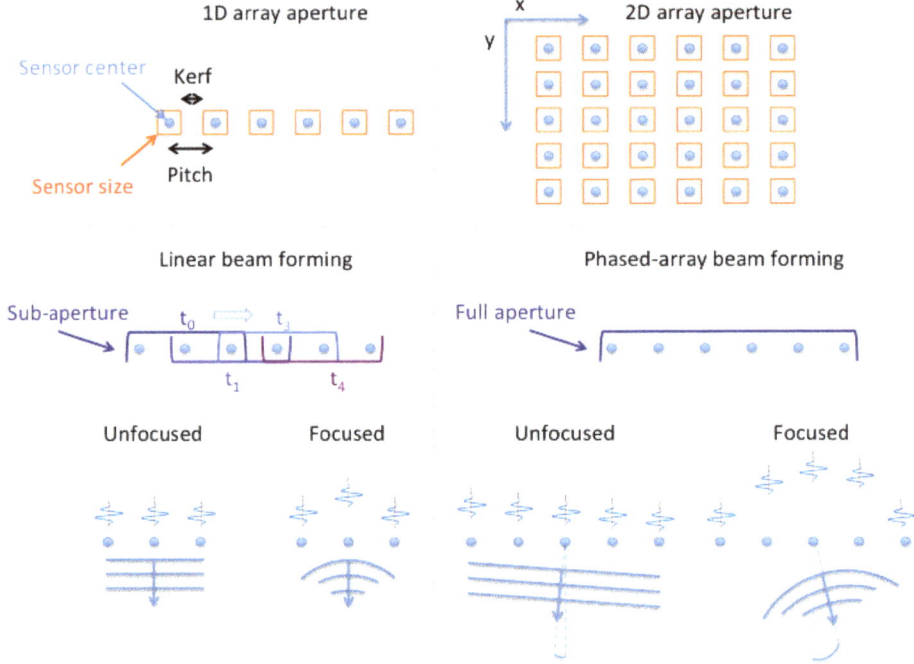

Figure 1. This figure shows the different sensor distributions for a one-dimensional (1D) and two-dimensional (2D) array aperture (top), together with an overview of the possible driving schemes for linear and phased array beam forming (bottom) in cases with a focused and an unfocused beam.

In principle, as described for point sources, each sensor can be excited by a signal having its own amplitude, phase, and waveform. However, sensors are generally grouped in sub-apertures, and within one sub-aperture, the same waveform is transmitted, but with a different phase and amplitude.

This is true for linear array beam forming, where a sub-aperture is defined and used both to transmit and receive ultrasound fields. The signal so acquired is then representative of the structures seen by the ultrasound waves over depth and in front of the sub-aperture. This signal is called an A-scan [2,3]. Subsequently, this sub-aperture is linearly shifted over the entire array so as to obtain multiple A-scans, ultimately forming an image line by line. The sensors that belong to a sub-aperture could be excited by signals that share the same phase, i.e., the unfocused case, or have different phases, as in the focused case. As can be deduced from Equation (4), when focusing is applied, higher pressures are generated. Moreover, smaller beams may even be achieved. Thus, focusing implies that the spatial (in the lateral and/or elevation direction) resolution, SNR, and penetration depth are improved. On the other hand, the area investigated by every beam is smaller, which means that more beams are necessary to cover a given FOV compared to the unfocused case. This also implies that more transmission events are required to form an image, which may decrease the frame rate. Within a given sub-aperture, the sensors could also be excited with different amplitudes. This is true if an apodization mask is used. Using an apodization mask reduces the amplitude of side lobes and their effects on the final image, but negatively affects the lateral and/or elevation spatial resolution. Furthermore, the maximum pressure generated is reduced, and thus consequently so are the SNR and penetration depths [2,3].

Unlike linear array beam forming, with phased array beam forming, the entire array aperture is used for each transmission. The phases of the driving signals are specifically adjusted for every sensor at each transmission event so as to steer the beam, and place it at a given angle with respect to the direction that is normal to the array aperture [2,3]. Different sets of phases are then used to obtain different steering directions, produce multiple A-scans, and thus form an image. With phased array beam forming, the beam could be a focused or unfocused beam, and apodization could be used. It is important to add that a particular constrain is present for phased array beam forming: the pitch has to be smaller than half the wavelength in order to avoid grating lobes. These are additional lobes, which can further degrade the image quality [2,3]. In Figure 1, a schematic overview of what has been introduced in this section is presented. Linear and phased array beam forming strategies are represented only for a 1D aperture, but these can of course be also applied to a 2D aperture, which gives more flexibility in the definition of the sub-apertures. Moreover, with a 2D aperture, the beam can be steered through the entire volume, rather than only on a plane perpendicular to the aperture [2,3].

When comparing linear and phased array beam forming, a list of pros and cons can be made. Both approaches form an image line by line, with one line being generated at every transmission event. Linear arrays can image only the area in front of the aperture, while a larger area can be imaged with phased arrays as the beam can be steered. This also means that the aperture of a linear array has to cover the entire area of interest (along the lateral direction). However, this is not the case for phased arrays. Consequently, phased arrays are particularly suitable in situations where there is a small imaging window, as in transthoracic ultrasound imaging, where the ribs represent an obstacle for imaging [5,6]. On the other hand, the geometries of phased arrays are constrained by the phenomenon of grating lobes, which is particularly demanding when using high frequencies. As a result, more accurate phase sets, and as many as the amount of steering angles, are required.

The transmit phase (or active phase), which is the phase that defines the shaping of the spatial distribution of the pressure field amplitude in the volume of interest, has been our focus thus far. In the receive phase, the very same phase sets and apodization functions that are used in the transmit phase can be applied. However, the received echo signals can be also treated differently. Furthermore, a different group of elements than those used in the transmission phase can also be used, as is, for example, the case for synthetic aperture beam forming [7,8] and multi-line acquisition beam forming [9]. Figure 2 illustrates the differences in the spatial distribution of the pressure amplitudes. The -20 dB beams obtained with a focused sub-aperture, and with an unfocused, focused, and steered full-aperture, are shown. The typical FOVs achievable with linear and phased array beam forming are also shown. Note that when the linear array sensors are distributed along a curve, a larger field of view can also be obtained. This is the case with convex probes. However, the probe loses its flat surface [10].

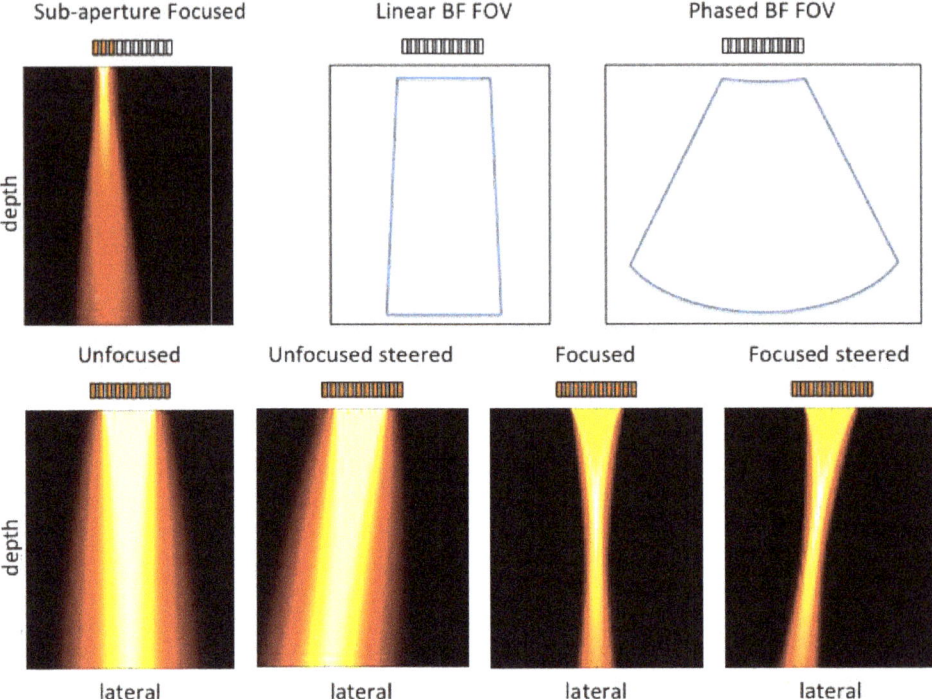

Figure 2. This figure shows the different spatial distribution of pressure amplitudes when unfocused, focused, and steered beams are generated. Moreover, the typical field of view (FOV) that is achievable with linear and phased array beam forming is also shown. These beam profiles were generated using the software package k-Wave [11].

3. Multi-Line Acquisition and Multi-Line Transmission Beam Forming

As briefly introduced above, the beam does not need to be the same in the transmit and receive phase. This is certainly the case with multi-line acquisition (MLA) beam forming. The basic idea behind this approach is to transmit a wide beam, so that a large area is covered, and then make use in receive of multiple, narrower beams, in order to form several A-scans along different directions for each transmission event. In this way, multiple lines are formed in parallel, thus increasing the frame rate and improving the temporal resolution. The receive phase is in fact defined by how the different signals received by all of the array elements are combined to form a line of the image. Therefore, it is possible to apply different phase sets and apodization masks to the signals received after a single transmission event, thus allowing the formation of multiple lines in parallel. In fact, these techniques are also referred to as parallel receive beam forming. In Equation (10), we can see that a wider beam can be achieved by using, without focusing, a small sub-aperture at the center of the array during transmission [9,12–14]. Since not all of the elements are utilized, and as a result the active aperture is reduced in transmit, the maximum pressure generated is consequently lower compared to the case where all of the elements are used. Furthermore, the spatial resolution (although not in the axial direction) is not as good, as focusing is applied only in the receive phase. Not only can MLA be applied to achieve a gain in the frame rate, it can also be applied to improve the SNR and contrast by simply averaging consecutive images obtained at a higher temporal resolution than with standard beam forming (i.e., techniques where only one line is generated per each transmission event).

Moreover, MLA techniques can also be used to image a larger FOV. In this case, the gain in acquisition rate is used to widen the area covered by the imaging system. To summarize with a simple example, in case 4, image lines could be formed in parallel, which means that: (a) the temporal resolution could be improved by factor of four, or (b) four consecutive images could be averaged to improve the SNR, or (c) a FOV that is four times larger could be in principle imaged, or (d) a combination of these gains could be achieved by spending the higher data acquisition rate in the most desirable way (e.g., averaging only two consecutive images and thus improving the SNR while still improving also the frame rate by factor of two).

A similar concept could be also applied at inverted phases: instead of having parallel lines being formed in the receive phase, they could be generated during transmission. This approach is referred to as multi-line transmission (MLT) or parallel transmit beam forming. Even in the case where the very same phase sets and apodization functions are used, and are simply swapped between the transmit phase and the receive phase, advantages can already be obtained. This is the case for tissue harmonic imaging applications. Unlike standard (fundamental) ultrasound imaging, this modality makes use of the harmonic components that are generated during ultrasound propagation, and not the pressure fields directly emitted by the array, to form an image. The harmonic components represent a part of the pressure wave fields, which is located around multiples of the transmitted center frequency. For a pulse-echo imaging system, an improved spatial resolution, a reduction of reverberation, grating, and side-lobe artifacts [15] are among the advantages of utilizing tissue harmonic imaging. In particular, MLT beam forming is better than MLA when applied to tissue harmonic imaging, because the higher pressure amplitude that is generated—thanks to a focused beam in transmission—is fundamental to boost the generation of sufficiently strong harmonic components. When applied to harmonic imaging, MLT beam forming provides a further reduction of the side-lobe amplitudes and an increase in SNR [16].

To generate multiple beams in transmission, different approaches are possible. One approach is to simply distribute multiple focused beams in the volume of interest. This is achieved by a linear superposition of the signals that are used to generate each individual beam. As a side effect, this limits the maximum signal strength that is applicable to the formation of every beam, and thus lowers the maximum pressure that can be generated by a single focused beam [17]. The MLT approach can be used both with linear array and phased array beam forming, and with 1D and 2D arrays [17–19]. To minimize the possible inter-beam interference generated by neighboring transmitted beams, specific sets of apodization functions can be applied [20]. Additionally, another approach to reduce the interbeam interference is to separate the different beams in the frequency domain. With this approach, which is referred to as frequency division multiplexing, the available transducer bandwidth is divided into orthogonal sub-bands, each of which is allocated to a beam. Multiple beams, as many as the number of sub-bands, can thus be transmitted in parallel, and the generated echo signals can then be identified in the receive phase by means of band-pass filters [21]. The main disadvantage of this method is the loss in axial resolution due to the subdivision of the available band into smaller sub-bands. A smaller sub-band implies, in fact, a longer pulse.

In general, when implementing MLT, it is beneficial to add small time delays between the signals that are used to generate multiple beams in transmission. This improves their capability to separate the different beams. As a side effect, it lengthens the transmit phase, and thus increases the depth for which the final imaging system will be blind [17,18]. With MLT, the inter-beam interference level required by the specific application of interest limits the number of parallel beams. In general, a higher number of parallel beams results in a higher level of interference [18–20].

Figure 3 illustrates the effect of inter-beam interference on the final image. An ultrasound image of four wire targets obtained by linear array beam forming is shown (top left corner). The same targets are then imaged using MLT applied to linear array beam forming and performed by frequency division multiplexing, with three beams in transmission (top right corner). As can be seen, four MLT "ghost" wires appear before and in front of each "actual" wire. This type of artifact is also present when MLT is performed by spatially distributing the transmitted beams over the volume of interest. However, the location of the artifact is different. To illustrate this phenomenon, a single wire is imaged using MLT without frequency division multiplexing with six beams in transmission (bottom). In this case, the image was obtained by MLT applied to phased array beam forming. Ghost wires are visible on the sides of the actual wire.

As for MLA, and also with MLT, the higher data acquisition rate achieved by generating multiple beams in transmission is not solely applicable to improve the frame rate. For example, when implementing MLT by means of orthogonal frequency division multiplexing, a multi-focusing imaging approach can be realized where the different sub-bands are used to generate beams with a focus at different depths. In particular, the lower the center frequency of the sub-band, the deeper the focus will be. Thanks to this approach, the penetration depth and signal-to-noise ratio (SNR) improves without affecting the frame rate [22]. In conclusion, it is important to note that MLA and MLT techniques can be combined together to have a multiplicative effect on the gain in the data acquisition rate [21,23].

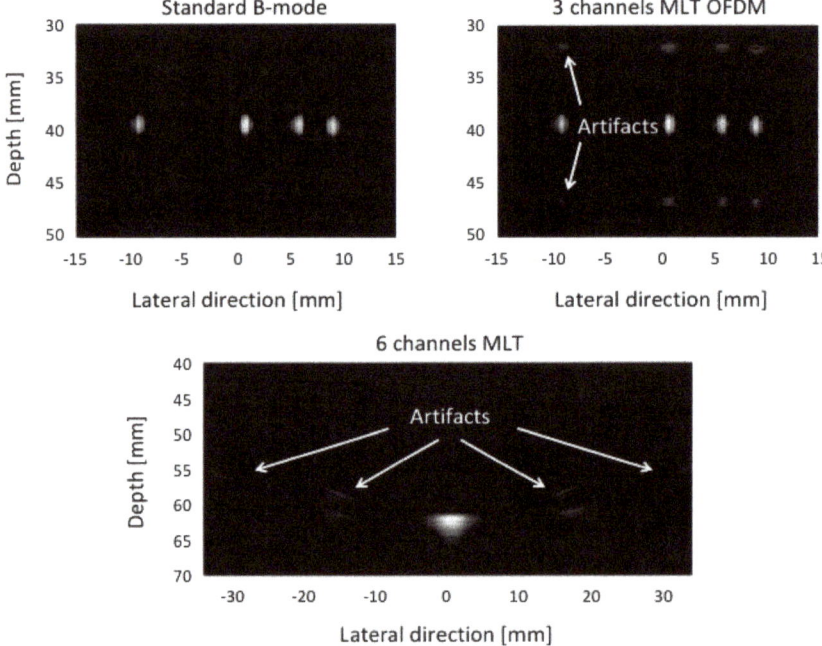

Figure 3. This figure illustrates the effect of inter-beam interference on the image. A standard linear array beam forming image of four wires is shown on the top left corner. The same wires are then imaged with multi-line transmission (MLT) performed by frequency division multiplexing, and applied to linear array beam forming. The corresponding image is shown in the top right corner, and "ghost" wires are clearly visible before and after each wire. At the bottom, an image of a single wire obtained with MLT performed without frequency division multiplexing, and applied to phased array beam forming, is shown. Also in this case, ghost wires are visible, but at a different position relative to the actual wire.

4. Plane and Diverging Wave Beam Forming

With these techniques, the emphasis is certainly more on improving the achievable frame rate. Plane wave imaging originates from studies aimed at imaging the transient propagation of shear mechanical waves in real time [24–26]. For this type of application, a frame rate in the order of thousands of frames per second is needed. The basic concept is that if one can reduce the number of transmission events that are needed to form an image to the bare minimum, this implies maximizing the frame rate. The absolute minimum is of course one transmission event per image. In this case, the frame rate is, in essence, only limited by the speed of the ultrasound wave in the imaging medium, the depths that need to be visualized, and the processing time that is necessary to form the image. Thus, the idea is to approximate in transmission the generation of a plane wave, achieving, in this way, a very wide and homogeneous beam (as wide as the array aperture). This can be achieved by simply exciting all of the transducer elements with the same phase for each transmission event. Then, in the receive phase, the signals acquired by all of the elements are processed with different phase sets and amplitudes, and multiple lines are generated in parallel. In particular, all of the lines that form an image are generated with the echo signals, which are received after a single transmission. In simple words, this approach can be seen as a radical MLA approach where only one large beam is used in transmission. As in the case of MLA, this technique suffers from lower pressure amplitudes being generated compared to focused beams, thus affecting the SNR and penetration depths. Moreover, the straightforward approach of plane wave imaging suffers from low image quality in terms of spatial resolution and contrast [27]. In fact, all of the imaging features are sacrificed to maximize the temporal resolution. A way to balance the performance of plane wave imaging among the different imaging features is to compromise, or in other words, to apply image compounding. Compounding essentially means averaging. However, it is not a simple averaging of consecutive frames. With plane wave coherent compounding, steering is applied, and thus the "plane wave" is no longer propagating only straight in front of the transducer array, it is also propagating under a given angle [27]. An image is then formed for varying transmission angles, and in the end, averaging is performed over the images obtained with all of the different angles. In this way, the gain in frame rate is reduced by a factor that is equal to the number of angles. On the other hand, the other imaging features (spatial resolution, SNR, penetration depth, contrast) are improved. However, in order to obtain a performance that is comparable to standard beam forming, the amount of compounded angles is very high (in the order of 70), and the extreme gain in frame rate that is achievable with plane wave imaging is substantially lost [27,28]. The number of angles, as well as the maximum steering angle, can be adjusted to tune plane wave imaging for a specific application. However, the most important feature of this technique is its capability to reach really high frame rates, which makes it extremely suitable for applications where fast phenomena need to be observed. In this situation, the spatial resolution is actually less important, while the key feature is the temporal resolution. Shear wave imaging is certainly a good example [29,30]. Other interesting areas of application are flow, contrast dynamics, and functional ultrasound imaging [31]. It is also important to mention that the implementation of this high frame rate imaging method has also been made possible thanks to the developments of GPU technologies, which provide the computational speed that is required to process the amount of data generated during plane wave imaging [32–34].

Diverging wave beam forming does not differ substantially from plane wave imaging. The small difference between the two methods lies in a defocused beam being used in transmission during diverging wave beam forming, which allows for an even larger insonification area [28,35–37]. Multiple parallel lines are generated in the receive phase in this case also, and compounding algorithms can be applied. In general, the same considerations as in plane wave imaging apply. Neither plane wave or diverging wave imaging are ideal for applications where a small array aperture is required, as is the case in transthoracic ultrasound imaging, where the presence of the ribs constrain the size of the imaging window. Moreover, due to the low-pressure amplitudes that are generated in the transmit phase, these techniques are certainly not ideal for tissue harmonic imaging applications, where high-pressure values are needed in order to generate the harmonic components that are necessary to form the image [16,38,39].

5. Synthetic Aperture Beam Forming

Synthetic aperture is a beam-forming approach that originates from the world of radar and was first implemented for medical ultrasound imaging in the late 1960s and early 1970s [40,41]. In its basic implementation, only one element is excited for every transmission event [42]. In the receive phase, all of the elements of the array are used to receive the echo signals, and a low quality image is generated for every transmission event. The key aspect is that every point of the image is obtained by taking into account the geometrical distance between each transmitting element and each receiving element. Thus, assuming a constant speed of sound through the imaging volume, appropriate phase sets are used to compensate for the differences in the arrival time. Subsequently, the received time-compensated signals are added together.

As a result, the images that are obtained for each transmission are combined to obtain an image of higher quality in terms of spatial resolution, contrast, and penetration depth with respect to the images obtained for every transmission. Thus, focusing is performed for every pixel in the image, and applied both in the transmit phase (indirectly by recombining the images formed with a single emitter) and the receive phase. As a consequence, the highest possible spatial resolution for delay-and-sum beam forming is obtained everywhere in the image [43]. However, the signal-to-noise ratio and penetration depths are significantly degraded by the array aperture being minimized in transmission, since only one element is active. Transmitting with sub-apertures rather than with a single element can mitigate this phenomenon [44–46]. However, the accuracy in the image reconstruction given by the availability of the data as obtained from the entire transmitting–receiving pairs of elements is lost, which deteriorates the spatial resolution. Once again, improving the performance with respect to a given imaging feature implies accepting that there will be a loss in performance with respect to another.

An interesting approach based on frequency division multiplexing has been proposed to improve the SNR and penetration depth without losing access to the full element-to-element data set [47]. Similarly to the case discussed for MLT, the available transducer bandwidth is divided into sub-bands. During each transmission, all of the elements are active, with each operating at one specific sub-band. During consecutive transmissions, every element is active at a different sub-band, and the entire bandwidth is covered. In the receive phase, band-pass filters are used to separate and identify the signal coming from the different elements. Using this approach, the entire aperture is active for every transmission event, and the achievable SNR and penetration depth are thus improved. Another possibility is to use chirp signals in transmission, and a matched filter in the receive phase. Particular attention to the signal properties is needed when chirp signals are used, so as to avoid temporal side lobes. Furthermore, additional processing steps and the ability of the hardware to generate well-controlled electrical signals is also required. However, this approach can improve penetration depth and axial resolution [48–50]. It is also important to note that if only one element is active for every transmission event, this implies that the time that is needed to collect all of the signals necessary to form an image is maximized, which in other words means minimizing the frame rate. However, not all of the elements of the array need to be used in transmission. In this way, a higher frame rate can be achieved. On the other hand, this will lower the spatial resolution and increase the amplitude of the side-lobes and their effect on the final image [51].

6. Comparison among Different Beam-Forming Options

A general comparison between the different techniques discussed thus far can be drawn. Figure 4 illustrates the different driving schemes for MLA, MLT, plane wave, diverging wave, and synthetic aperture beam forming. The transmit beams are represented in orange, and different shades of orange are used to highlight the multiple beams for MLT, and separate the beam profiles for plane and diverging wave beam forming, respectively. For MLA, the receive beams are also shown in shades of blue. The duration of the transmit phase is also emphasized for MLT. Table 1 summarizes the peculiarities of each modality. A comparison is made with standard line-by-line beam forming. A plus or minus sign means that the performance with respect to that specific imaging feature (one for each column) is improved or reduced, respectively. MLA and MLT beam forming is essentially trading spatial resolution for data acquisition rate, and can be generally applied to any array aperture. In the case of MLT especially, where focused beams are used in transmission, penetration depth is not lost compared to standard beam forming. Plane and diverging wave beam forming focus instead on achieving a very high data acquisition rate. Consequently, spatial resolution and penetration depth are affected. Moreover, these approaches substantially require a large aperture size. This is also true for synthetic aperture beam forming, where an increase in the number of transmitting elements leads to improved performance. The strength of synthetic aperture beam forming is certainly on the attainable spatial resolution, which is achieved at the expense of penetration depth and the data acquisition rate.

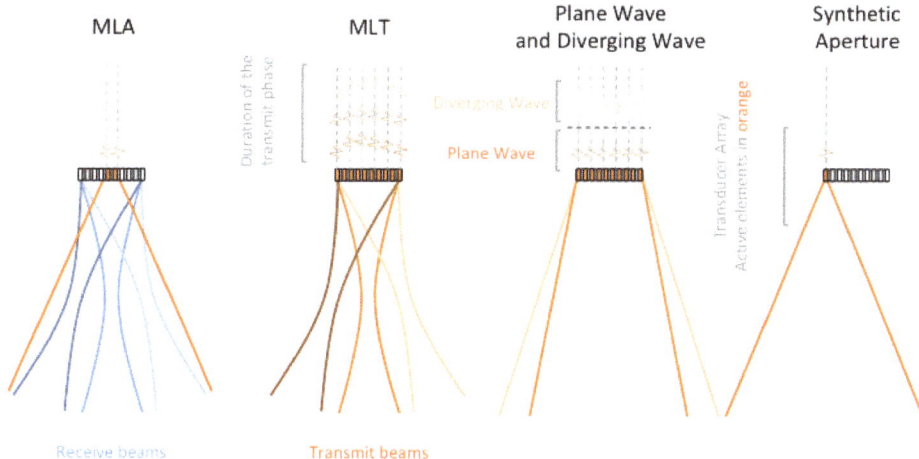

Figure 4. This figure illustrates the different driving schemes for MLA, multi-line transmission (MLT), plane wave, diverging wave, and synthetic aperture beam forming. The transmit beams are represented in orange, and different shades of orange are used to highlight multiple beams for MLT, and separate the beam profiles for plane and diverging wave, respectively. For MLA, the receive beams are also shown in shades of blue. The duration of the transmit phase is also emphasized for MLT.

As described in the previous sections, for every technique, these pros and cons can be mitigated by specific implementations. However, the rule that what one gains regarding particular feature implies a loss in performance with respect to the other features generally applies.

Table 1. This table summarizes the peculiarities of each modality. A plus or minus sign means that the performance with respect to that specific imaging feature (one for each column) is improved or reduced, respectively. The 0 sign means that there is no significant variation. The evaluation is performed with respect to standard beam-forming performance.

Beam Forming Strategy	Spatial Resolution	Data Acquisition Rate	Array Aperture Size	Penetration Depths
MLA–MLT	−	+	0	0
Plane and Diverging Wave	−	++	−	−
Synthetic Aperture	+	−	−	−

7. Other Beam-Forming Strategies

Several emerging beam forming strategies, besides those dealt with in this paper, have been reported and discussed in the literature, including approaches based on machine learning [52]. Particularly interesting concepts are those explored with null subtraction imaging (NSI), and coherence beam forming. With NSI, particular sets of apodization functions are used to achieve a lateral (and potentially elevation) spatial resolution that goes beyond the diffraction limit. This technique requires the application of signal-processing techniques only in the receive phase. In essence, the idea is to combine the images formed using zero mean and non-zero mean apodization functions. A zero-mean apodization generates a beam with a "hole" along the beam axis (see Huygens' principle). Consequently, this beam can be subtracted from that generated by a non-zero mean apodization function, thus obtaining an extremely sharp beam. However, the gain in spatial resolution is costly in terms of contrast [53–55].

When using coherence beam forming, each imaging pixel is obtained from the integration of the normalized covariance matrix that is calculated between the signals received by all of the elements forming the array [56]. This follows after appropriate time-delay compensations are applied. This technique is clearly more computationally expensive compared to standard beam forming, and generally results in a smaller dynamic range. However, it is particularly suitable for applications in low SNR imaging conditions. Contrast is generally improved, and noise is significantly reduced [56–58].

8. Conclusions

In this paper, a review of different beam-forming schemes has been presented. Multi-line acquisition, multi-line transmission, plane wave, diverging wave, synthetic aperture, and more recent beam-forming strategies have been introduced. The peculiarities and advantages of these approaches have been compared, and their applicability has been discussed. In conclusion, it is also important to mention that all of these studies would have not been possible without the development of open research platforms, thanks to which the implementation and testing of advanced beam-forming strategies were carried out [59–64].

Funding: This research received no external funding.

Conflicts of Interest: The authors declare no conflict of interest.

References

1. Demi, L. Modeling Nonlinear Propagation of Ultrasound through Inhomogeneous Biomedical Media. Ph.D. Thesis, Delft University of Technology, Delft, The Netherlands, 2013.
2. Szabo, T. *Diagnostic Ultrasound Imaging: Inside Out*, 2nd ed.; Elsevier: New York, NY, USA, 2013.
3. Cobbolt, R.S.C. *Foundation of Biomedical Ultrasound*; Oxford University Press: Oxford, UK, 2006.
4. Ramalli, A.; Boni, E.; Savona, A.S.; Tortoli, P. Density-tapered spiral arrays for ultrasound 3-D imaging. *IEEE Trans. Ultrason. Ferroelectr. Freq. Control* **2015**, *8*, 1580–1588. [CrossRef] [PubMed]
5. Ramm, O.T.; Thurstone, F.L. Cardiac imaging using a phased array ultrasound system I System Design. *Circulation* **1976**, *2*, 258–262.
6. Ramm, O.T.; Thurstone, F.L. Cardiac imaging using a phased array ultrasound system II Clinical Techniques and Applications. *Circulation* **1976**, *2*, 262–267.
7. Norton, S.J. Acoustical holography with an annular aperture. *J. Acoust. Soc. Am.* **1982**, *71*, 1169–1178. [CrossRef]
8. Ajay, K.L.; Kassam, S.A. Side-lobe reduction in the ring array pattern for synthetic aperture imaging of coherent sources. *J. Acoust. Soc. Am.* **1983**, *74*, 840–846.
9. Shattuck, D.P.; Weinshneker, M.D. Explososcan: A parallel processing technique for high speed ultrasound imaging with linear phased arrays. *J. Acoust. Soc. Am.* **1984**, *75*, 1273–1282. [CrossRef] [PubMed]
10. Szabo, T.L.; Lewin, P.A. Ultrasound transducer selection in clinical imaging practice. *J. Ultrasound Med.* **2013**, *32*, 573–582. [CrossRef] [PubMed]
11. Treeby, B.E.; Cox, B.T. K-Wave: MATLAB toolbox for the simulation and reconstruction of photoacoustic wave-fields. *J. Biomed. Opt.* **2010**, *15*, 021314. [CrossRef] [PubMed]
12. Smith, S.W.; Pavy, H.G., Jr.; von Ramm, O.T. High Speed Ultrasound Volumetric Imaging System Part I Transducer Design and Beam Steering. *IEEE Trans. Ultrason. Ferroelectr. Freq. Control* **1991**, *38*, 100–108. [CrossRef] [PubMed]

13. Von Ramm, O.T.; Smith, S.W.; Pavy, H.G., Jr. High Speed Ultrasound Volumetric Imaging System Part II parallel processing and image display. *IEEE Trans. Ultrason. Ferroelectr. Freq. Control* **1991**, *38*, 109–114. [CrossRef] [PubMed]
14. Snyder, J.E.; Kisslo, J.A.; von Ramm, O.T. Real time orthogonal mode scanning of the heart: A new ultrasonic imaging modality. *J. Am. Coll. Cardiol.* **1986**, *7*, 1279. [CrossRef]
15. Demi, L.; Verweij, M.D. Nonlinear Acoustics. In *Comprehensive Biomedical Physics*; Elsevier: New York, NY, USA, 2014.
16. Demi, L.; Verweij, M.D.; van Dongen, K.W.A. Parallel Transmit Beamforming Using Orthogonal Frequency Division Multiplexing and Applied to Harmonic Imaging—A Feasibility Study. *IEEE Trans. Ultrason. Ferroelectr. Freq. Control* **2012**, *59*, 2439–2447. [CrossRef] [PubMed]
17. Santos, P.; Tong, L.; Ortega, A.; Løvstakken, L.; Samset, E.; D'hooge, J. Safety of Multi-Line Transmit beam forming for fast cardiac imaging—A simulation study. In Proceedings of the IEEE International Ultrasonics Symposium, Chicago, IL, USA, 3–6 September 2014; pp. 1199–1202.
18. Demi, L.; Viti, J.; Kusters, L.; Guidi, F.; Tortoli, P.; Mischi, M. Implementation of Parallel Transmit Beamforming Using Orthogonal Division Multiplexing—Achievable Resolution and Interbeam Interference. *IEEE Trans. Ultrason. Ferroelectr. Freq. Control* **2013**, *60*, 2310–2320. [CrossRef] [PubMed]
19. Denarie, B.; Bjastad, T.; Torp, H. MemberMulti-line Transmission in 3D with Reduced Cross-talk Artifacts—A Proof of Concept Study. *IEEE Trans. Ultrason. Ferroelectr. Freq. Control* **2013**, *60*, 1708–1718. [CrossRef] [PubMed]
20. Tong, L.; Gao, H.; D'hooge, J. Multi-Transmit Beamforming for Fast Cardiac Imaging—A Simulations Study. *IEEE Trans. Ultrason. Ferroelectr. Freq. Control* **2013**, *60*, 1719–1731. [CrossRef] [PubMed]
21. Demi, L.; Ramalli, A.; Giannini, G.; Mischi, M. In Vitro and In Vito Tissue Harmonic Imaging Obtained with Parallel Transmit Beamforming by Means of Orthogonal Frequency Division Multiplexing. *IEEE Trans. Ultrason. Ferroelectr. Freq. Control* **2015**, *62*, 2030–2035. [CrossRef] [PubMed]
22. Demi, L.; Giannini, G.; Ramalli, A.; Tortoli, P.; Mischi, M. Multi-focus Tissue Harmonic Imaging Obtained with Parallel Transmit Beamforming by Means of Orthogonal Frequency Division Multiplexing. In Proceedings of the IEEE International Ultrasonics Symposium, Taipei, Taiwan, 21–24 October 2015.
23. Tong, L.; Ramalli, A.; Jasaityte, R.; Tortoli, P.; D'hooge, J. Multi-Transmit Beam forming for Fast Cardiac Imaging—Experimental Validation and In Vivo Application. *IEEE Trans. Med. Imaging* **2015**, *33*, 1205–1219. [CrossRef] [PubMed]
24. Sandrin, L.; Catheline, S.; Tanter, M.; Hennequin, X.; Fink, M. Time resolved pulsed elastography with ultrafast ultrasonic imaging. *Ultrason. Imaging* **1999**, *21*, 259–272. [CrossRef] [PubMed]
25. Sandrin, L.; Catheline, S.; Tanter, M.; Hennequin, X.; Vinconneau, C.; Fink, M. 2D transient elastography. *Acoust. Imaging* **2000**, *25*, 485–492.
26. Sandrin, L.; Catheline, S.; Tanter, M.; Fink, M. Shear modulus imaging using 2D transient elastography. *IEEE Trans. Ultrason. Ferroelectr. Freq. Control* **2002**, *49*, 426–435. [CrossRef] [PubMed]
27. Montaldo, G.; Tanter, M.; Bercoff, J.; Benech, N.; Fink, M. Coherent Plane-Wave Compounding for Very High Frame Rate Ultrasonography and Transient Elastography. *IEEE Trans. Ultrason. Ferroelectr. Freq. Control* **2009**, *56*, 489–506. [CrossRef] [PubMed]
28. Tong, L.; Gao, H.; Fai Choi, H.; D'hooge, J. Comparison of Conventional Parallel Beam Forming with Plane Wave and Diverging Wave Imaging for Cardiac Applications—A Simulation Study. *IEEE Trans. Ultrason. Ferroelectr. Freq. Control* **2012**, *59*, 1654–1663. [CrossRef] [PubMed]
29. Bercoff, J.; Tanter, M.; Fink, M. Sonic boom in soft materials: The elastic Cerenkov Effect. *Appl. Phys. Lett.* **2004**, *84*, 2202–2204. [CrossRef]
30. Bercoff, J.; Tanter, M.; Fink, M. Supersonic shear imaging: A new technique for soft tissues elasticity mapping. *IEEE Trans. Ultrason. Ferroelectr. Freq. Control* **2004**, *51*, 396–409. [CrossRef] [PubMed]
31. Tanter, M.; Fink, M. Ultrafast Imaging in Biomedical Ultrasound. *IEEE Trans. Ultrason. Ferroelectr. Freq. Control* **2014**, *61*, 102–119. [CrossRef] [PubMed]
32. So, H.; Chen, J.; Yiu, B.; Yu, A. Medical ultrasound imaging: To GPU or not to GPU? *IEEE Micro* **2011**, *31*, 54–65. [CrossRef]
33. Yiu, B.Y.S.; Tsang, I.K.H.; Yu, A.C.H. GPU-based beamformer: Fast realization of plane wave compounding and synthetic aperture imaging. *IEEE Trans. Ultrason. Ferroelectr. Freq. Control* **2011**, *58*, 1698–1705. [CrossRef] [PubMed]

34. Martin-Arguedas, C.J.; Romero-Laorden, D.; Martinez-Graullera, O.; Perez-Lopez, M.; Gomez-Ullate, L. An ultrasonic imaging system based on a new SAFT approach and a GPU beamformer. *IEEE Trans. Ultrason. Ferroelectr. Freq. Control* **2012**, *59*, 1402–1412. [CrossRef] [PubMed]
35. Hasegawa, H.; Kanai, H. High-frame-rate Echocardiography Using diverging transmit beams and parallel receive beamforming. *J. Med. Ultrason.* **2011**, *38*, 129–140. [CrossRef] [PubMed]
36. Papadacci, C.; Pernot, M.; Couade, M.; Fink, M.; Tanter, M. High contrast ultrafast imaging of the heart. *IEEE Trans. Ultrason. Ferroelectr. Freq. Control* **2014**, *61*, 288–301. [CrossRef] [PubMed]
37. Santos, P.; Haugen, G.U.; Løvstakken, L.; Samset, E.; D'hooge, J. Diverging Wave Volumentric Imaging Using Sub-aperture Beam forming. *IEEE Trans. Ultrason. Ferroelectr. Freq. Control* **2016**, *63*, 2114–2124. [CrossRef] [PubMed]
38. Cikes, M.; Tong, L.; Sutherland, G.R.; D'hooge, J. Ultrafast cardiac ultrasound imaging, Technical Principles, Applications and Clinical Benefits. *JACC Cardiovasc. Imaging* **2014**, *7*, 812–823. [CrossRef] [PubMed]
39. Prieur, F.; Denarie, B.; Austeng, A.; Torp, H. Multi-line transmission in medical imaging using the second-harmonic signal. *IEEE Trans. Ultrason. Ferroelectr. Freq. Control* **2013**, *60*, 2682–2692. [CrossRef] [PubMed]
40. Flaherty, J.J.; Erikson, K.R.; Lund, V.M. Synthetic Aperture Ultrasound Imaging Systems. U.S. Patent 3,548,642, 2 March 1967.
41. Burckhardt, C.B.; Grandchamp, P.A.; Hoffmann, H. An experimental 2 MHz synthetic aperture sonar system intended for medical use. *IEEE Trans. Ultrason. Ferroelectr. Freq. Control* **1974**, *21*, 1–6. [CrossRef]
42. Ylitalo, J.T.; Ermert, H. Ultrasound Syntheric Aperture Imaging: Monostatic approach. *IEEE Trans. Ultrason. Ferroelectr. Freq. Control* **1994**, *41*, 333–339. [CrossRef]
43. Jensen, J.A.; Nikolov, S.I.; Gammelmark, L.; Pedersen, M.H. Synthetic Aperture Ultrasound Imaging. *Ultrasonics* **2006**, *44*, e5–e15. [CrossRef] [PubMed]
44. O'Donnel, M.; Thomas, L.J. Efficient Synthetic Aperture Imaging from a circular aperture with possible application to catheter-based imaging. *IEEE Trans. Ultrason. Ferroelectr. Freq. Control* **1992**, *39*, 366–380. [CrossRef] [PubMed]
45. Karaman, M.; Li, P.C.; O'Donnel, M. Synthetic aperture imaging for small scale systems. *IEEE Trans. Ultrason. Ferroelectr. Freq. Control* **1995**, *42*, 429–442. [CrossRef]
46. Karaman, M.; O'Donnel, M. Subaperture processing for ultrasonic imaging. *IEEE Trans. Ultrason. Ferroelectr. Freq. Control* **1998**, *45*, 126–135. [CrossRef] [PubMed]
47. Fredrik, G.; Jensen, J.A. Frequency Division Transmission Imaging and Synthetic Aperture Reconstruction. *IEEE Trans. Ultrason. Ferroelectr. Freq. Control* **2006**, *53*, 900–911.
48. Misaridis, T.X.; Gammelmark, K.; Jorgensen, C.H.; Lindberg, N.; Thomsen, A.H.; Pedersen, M.H.; Jensen, J.A. Potential of coded excitation in medical ultrasound imaging. *Ultrasonics* **2000**, *38*, 183–189. [CrossRef]
49. Misaridis, T.; Jensen, J.A. Use of modulated excitation signals inultrasound. Part I: Basic concepts and expected benefits. *IEEE Trans. Ultrason. Ferroelectr. Freq. Control* **2005**, *52*, 192–207. [CrossRef] [PubMed]
50. Gammelmark, K.L.; Jensen, J.A. Multielement synthetic transmit aperture imaging using temporal encoding. *IEEE Trans. Med. Imaging* **2003**, *22*, 552–563. [CrossRef] [PubMed]
51. Nikolov, S.I. Synthetic Aperture Tissue and Flow Ultrasound Imaging. Ph.D. Thesis, DTU Technical University of Denmark, Lyngby, Denmark, 2001.
52. Luchies, A.; Byram, B.C. Deep Neural Networks for Ultrasound Beamforming. *IEEE Trans. Med. Imaging* **2018**, *37*, 1464–1477. [CrossRef] [PubMed]
53. Savoia, A.S.; Matrone, G.; Ramalli, A.; Boni, E. Improved lateral resolution and contrast in ultrasound imaging using sidelobe masking technique. In Proceedings of the IEEE International Ultrasonics Symposium, Chicago, IL, USA, 3–6 September 2014.
54. Reeg, J.R.; Oelze, M.L. Improving lateral resolution in ultrasonic imaging by utilizing nulls in the beam pattern. In Proceedings of the IEEE International Ultrasonics Symposium, Taipei, Taiwan, 21–24 October 2015.
55. Reeg, J.R. Null Subtraction Imaging Technique for Biomedical Ultrasound Imaging. Master's Thesis, Electrical and Computer Engineering, Graduate College of the University of Illinois Urbana-Champaign, Champaign, IL, USA, 2016.
56. Lediju, M.; Trahey, G.E.; Byram, B.C.; Dahl, J.J. Short-lag spatial coherence of backscattered echoes: Imaging characteristics. *IEEE Trans. Ultrason. Ferroelectr. Freq. Control* **2011**, *58*, 1377–1388. [CrossRef] [PubMed]

57. Dahl, J.J.; Hyun, D.; Lediju, M.; Trahey, G.E. Lesion detectability in diagnostic ultrasound with short-lag spatial coherence imaging. *Ultrason. Imaging* **2011**, *33*, 119–133. [CrossRef] [PubMed]
58. Dahl, J.J.; Hyun, D.; Li, Y.; Jakovljevic, M.; Bell, M.A.L.; Long, W.J.; Bottenus, N.; Kakkad, V.; Trahey, G.E. Coherence Beamforming and Its Applications to the Difficult-to-Image Patient. In Proceedings of the IEEE International Ultrasonics Symposium, Washington, DC, USA, 6–9 September 2017.
59. Jensen, J.A.; Holm, O.; Jensen, L.J.; Bendsen, H.; Nikolov, S.I.; Tomov, B.G.; Munk, P.; Hansen, M.; Salomonsen, K.; Hansen, J.; et al. Ultrasound research scanner for real-time synthetic aperture data acquisition. *IEEE Trans. Ultrason. Ferroelectr. Freq. Control* **2005**, *52*, 881–891. [CrossRef] [PubMed]
60. Jensen, J.A.; Holten-Lund, H.; Nilsson, R.T.; Hansen, M.; Larsen, U.D.; Domsten, R.P.; Tomov, B.G.; Stuart, M.B.; Nikolov, S.I.; Pihl, M.J.; et al. SARUS: A synthetic aperture real-time ultrasound system. *IEEE Trans. Ultrason. Ferroelectr. Freq. Control* **2013**, *60*, 1838–1852. [CrossRef] [PubMed]
61. Tortoli, P.; Bassi, L.; Boni, E.; Dallai, A.; Guidi, F.; Ricci, S. ULA-OP: An advanced open platform for ultrasound research. *IEEE Trans. Ultrason. Ferroelectr. Freq. Control* **2009**, *56*, 2207–2216. [CrossRef] [PubMed]
62. Boni, E.; Bassi, L.; Dallai, A.; Guidi, F.; Meacci, V.; Ramalli, A.; Ricci, S.; Tortoli, P. ULA-OP 256: A 256-channel open scanner for development and real-time implementation of new ultrasound methods. *IEEE Trans. Ultrason. Ferroelectr. Freq. Control* **2016**, *63*, 1488–1495. [CrossRef] [PubMed]
63. Lewandowski, M.; Walczak, M.; Witek, B.; Kulesza, P.; Sielewicz, K. Modular & scalable ultrasound platform with GPU processing. In Proceedings of the 2012 IEEE International Ultrasonics Symposium, Dresden, Germany, 7–10 October 2012.
64. Cheung, C.; Yu, A.; Salimi, N.; Yiu, B.; Tsang, I.; Kerby, B.; Azar, R.; Dickie, K. Multi-channel pre-beamformed data acquisition system for research on advanced ultrasound imaging methods. *IEEE Trans. Ultrason. Ferroelectr. Freq. Control* **2012**, *59*, 243–253. [CrossRef] [PubMed]

© 2018 by the author. Licensee MDPI, Basel, Switzerland. This article is an open access article distributed under the terms and conditions of the Creative Commons Attribution (CC BY) license (http://creativecommons.org/licenses/by/4.0/).

Article

Methods for Grating Lobe Suppression in Ultrasound Plane Wave Imaging

Sua Bae and Tai-Kyong Song *

Department of Electronic Engineering, Sogang University, Seoul 04107, Korea; suabae@sogang.ac.kr
* Correspondence: tksong@sogang.ac.kr; Tel.: +82-2-705-8907

Received: 29 August 2018; Accepted: 9 October 2018; Published: 11 October 2018

Abstract: Plane wave imaging has been proven to provide transmit beams with a narrow and uniform beam width throughout the imaging depth. The transmit beam pattern, however, exhibits strong grating lobes that have to be suppressed by a tightly focused receive beam pattern. In this paper, we present the conditions of grating lobe occurrence by analyzing the synthetic transmit beam pattern. Based on the analysis, the threshold of the angle interval is presented to completely eliminate grating lobe problems when using uniformly distributed plane wave angles. However, this threshold requires a very small angle interval (or, equivalently, too many angles). We propose the use of non-uniform plane wave angles to disperse the grating lobes in the spatial domain. In this paper, we present an approach using two uniform angle sets with different intervals to generate a non-uniform angle set. The proposed methods were verified by continuous-wave transmit beam patterns and broad-band 2D point spread functions obtained by computer simulations.

Keywords: ultrasonic imaging; beamforming; plane wave imaging; grating lobe suppression

1. Introduction

Plane wave imaging (PWI) has drawn a large amount of attention from researchers in the field of medical ultrasound imaging [1–3]. First, PWI can provide ultra-fast ultrasound imaging that is essential for a growing number of applications, such as the estimation of shear elasticity [1,4,5] and vector Doppler [6,7] as well as high-frame-rate B-mode imaging [8]. In PWI, plane waves (PWs) with different travelling angles are successively transmitted instead of traditionally focused ultrasound waves; after each firing, the returned ultrasound waves are received at all array elements. Synthetic transmit (Tx) focusing at each imaging point is achieved by compounding PWs with proper delays, while receive (Rx) focusing is performed in the conventional manner. As a result, ultrasound beams are focused at all imaging points for transmission and reception. Theoretically, the Tx beam pattern of PWI maintains the same main lobe width at all depths; the width is determined by the range of compounded PW angles.

When using a finite number of PWs with uniformly distributed steering angles, the synthetically focused beam has not only the main lobe, but also side lobes and grating lobes (GLs), which create artifacts in ultrasound image and deteriorate the image quality [2]. A large number of compounded PWs allow for the mitigation of the side lobe and GLs in the synthetic beam pattern and provide better image contrast, as illustrated in Figure 1. However, the frame rate of PWI decreases as the number of PWs increases. To reduce the side lobe without compromising the frame rate, various adaptive beamforming methods have been proposed. Austeng et al. proposed a minimum variance beamforming method for PWI [9]. In this method, the optimized weighting factors are applied when compounding the low-resolution images of different steered PWs. The joint Tx and Rx adaptive beamformer has also been proposed to apply the data-dependent weighting factors to both the receiving array domain and PW angle domain (i.e., frame domain) [10]. In addition, as the side lobes from different angles are uncorrelated, some beamformers have been suggested to reject less coherent signals [11–13]. A global effective distance-based side lobe suppressing method has also been proposed to achieve high quality images of PWI with a small number of PWs [14].

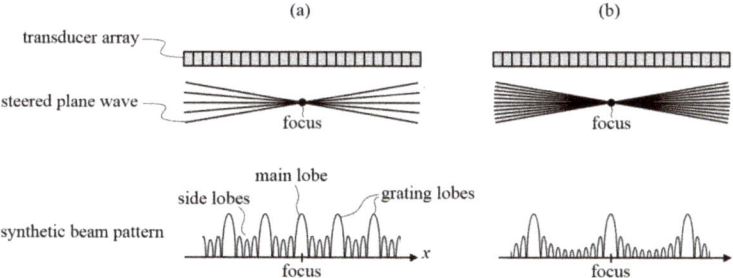

Figure 1. Schematic diagrams of a linear transducer array, transmitted plane waves (PWs) with uniformly distributed steering angles, and the synthetic transmit (Tx) beam pattern when (**a**) five PWs with a larger steering angle interval and (**b**) eleven PWs with a smaller angle interval are compounded. As the number of compounded PWs increases when the total range of the steering angle is fixed, the side lobe level decreases and grating lobes (GLs) occur less frequently.

Only a few studies, however, have been reported that consider GLs in PWI. Though PW angles with a constant angle interval are employed in most studies, they introduce uniformly spaced grating lobes (GLs), the interval of which is governed by the angle interval, as shown in Figure 1 [2]. If the angle interval is sufficiently small to locate GLs far from the main lobe, the GLs might have little effect on the image quality. However, to preserve the ultra-fast frame rate without compromising the resolution (i.e., to use a small number of PWs for the given PW angle range), the angle interval should be large, which introduces GLs close to the main lobe and deteriorates the image quality.

Here, we investigate the conditions of GL occurrence by analyzing the continuous wave (CW) synthetic transmit beam pattern. Based on the conditions of GL occurrence, the threshold of the angle interval for the elimination of GLs is presented with the use of uniformly distributed PW angles. In addition, we propose a method for GL level reduction using non-uniformly distributed PW angles, which consist of subsets of uniformly distributed angles, each of which has different angle intervals. To verify and evaluate the methods, simulation experiments are conducted using synthetic Tx beam patterns and round-trip point spread functions (PSFs).

2. Materials and Methods

2.1. GL Conditions in PWI

Let us consider N PWs with different steering angles, θ_n ($n = 1, 2, \ldots, N$), in a range of $[\theta_{\min}, \theta_{\max}]$, which are selected in terms of $\alpha_n (= \sin(\theta_n))$ for the convenience of analysis such that $\alpha_n = \alpha_{\min} + (n-1)d_\alpha$, $n = 1, 2, \ldots, N$, where $\alpha_{\min} = -(N-1)d_\alpha/2$ and d_α is a constant α-interval between successive α_n. Assuming that the monochromatic (i.e., CW) plane waves are emitted from an infinite-length transducer, PWI provides a synthetic Tx beam pattern given by:

$$\psi(x') = \sum_{n=1}^{N} \exp\{-jkx'(\alpha_{\min} + (n-1)d_\alpha)\} = c_0 \frac{\sin(\pi x' d_\alpha N/\lambda)}{\sin(\pi x' d_\alpha/\lambda)}, \qquad (1)$$

where $c_0 = \exp\{-jkx'(\alpha_{\min} + (N-1)d_\alpha/2)\}$, $x' = x - x_f$, x_f is the lateral position of a focus, λ is the wavelength, and k is the wave number ($k = 2\pi/\lambda$) [3]. Note that almost the same beam pattern can be obtained from Equation (1) when using PW angles spaced by a constant θ-interval, as in [1,2], if the angles are sufficiently small such that $\sin \theta_n \approx \theta_n$ and $d_\alpha \approx d_\theta$. In the beam pattern, the main lobe is at the focus, $x' = 0$ ($x = x_f$), and its amplitude is N. GLs must be observed for which both the numerator and denominator of Equation (1) have zeros except $x' = 0$, that occur at:

$$x' = \pm m\lambda/d_\alpha, m = 1, 2, 3, \ldots. \tag{2}$$

By substituting Equation (2) into Equation (1), the amplitude of the m-th GL can be expressed as:

$$\psi(x' = \pm m\lambda/d_\alpha) = \begin{cases} (-1)^m N & \text{for even } N \\ N & \text{for odd } N \end{cases}. \tag{3}$$

It should be noted that the GLs predicted by Equations (2) and (3) arise when (1) all of the N PWs pass through the GL positions, preserving their linear wave-front (LWF) (LWF condition), and (2) PWs with a constant α-interval are employed (uniform d_α condition).

2.2. GL Suppression Method with Uniformly Distributed PW Angles

In practice, PWs are transmitted by a finite aperture (i.e., a finite-length transducer array), resulting in a finite collimated beam area. Figure 2 shows three PWs with different steering angles (black solid lines) and their collimated beam areas (gray shaded areas) with preserved LWFs. In Figure 2, the focus and the main lobe of the focused beam pattern are in the region where all the PWs preserve their LWFs (i.e., the darkest area in Figure 2). If a GL locates at this same region and the constant PW angle interval is employed (i.e., if both LWF and uniform d_α conditions are met), the amplitude of the GL would be as high as that of the main lobe because the number of compounded PWs should be the same at both the main lobe and GL locations. Note that by reducing d_α, one can move the GL away from the main lobe towards a region where the LWF condition is satisfied by fewer PWs. In such a case, the GL level would be lower than that of the main lobe because the number of PWs coherently compounded becomes smaller at the GL location; the PWs that do not maintain LWFs at the grating lobe locations are compounded with phase errors, leading to a decrease in the corresponding GL levels. This indicates that GL levels would decrease as GLs are moved farther from the main lobe.

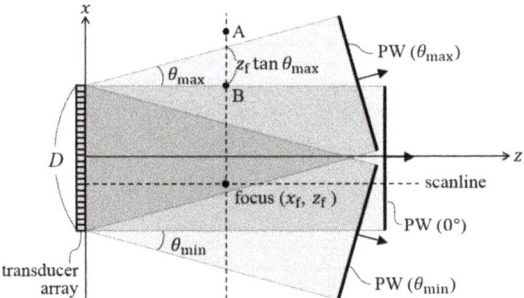

Figure 2. Collimated beam areas (gray shaded areas) of PWs with steering angles of θ_{max}, $0°$, and θ_{min} (from top to bottom). The points, A and B, are located at the region where none of PWs pass and the region where half of transmitted PWs propagate, respectively.

One can expect that the GLs can be eliminated by locating all of the GLs in a region where none of the PWs preserve LWF (i.e., by completely violating the LWF condition). In Figure 2, for example, when the first GL position, $x = x_{GL,1} (= x_f + \lambda/d_\alpha)$, falls onto point A that is out of all the collimated beam areas of PWs, no GLs can be formed, even when a uniform d_α is employed. In this case, $x_{GL,1} > D/2 + z_f \tan \theta_{max}$. Consequently, the GL elimination requirement when using uniformly distributed PW angles can be defined as:

$$d_\alpha < \lambda/(D/2 + z_f \tan \theta_{max} - x_f), \tag{4}$$

where (x_f, z_f) represents the focal point and D is the transducer width. One can also observe that only the PWs with $\theta \geq 0°$ pass through point B in Figure 2, preserving the LWF. Therefore, the GL level would be approximately halved at this point (i.e., reduced by -6 dB) if the angular interval is:

$$d_\alpha = \lambda/(D/2 - x_f), \tag{5}$$

when $\theta_{max} = -\theta_{min}$.

2.3. GL Suppression Method with Non-Uniformly Distributed PW Angles

The GL levels can also be reduced using non-uniformly distributed PW angles (i.e., by violating the uniform d_α condition). In this paper, to generate a non-uniform angle set, an approach using two uniform angle sets with different d_α values is presented.

We let the uniform angle set 1 and set 2 have N_1 PWs with an interval of $d_{\alpha,1}$ and N_2 PWs with an interval of $d_{\alpha,2}$, respectively. The GLs of set 1 would appear at integer multiples of $\lambda/d_{\alpha,1}$, while the GLs of set 2 would arise at integer multiples of $\lambda/d_{\alpha,2}$, according to Equation (2). To suppress the GL level, a non-uniform PW angle set can be obtained by combining the two uniform angle sets. The beam pattern of the non-uniform angle set is given by the sum of the beam patterns of the two uniform angle sets. Note that the main lobes of sets 1 and 2 are both centered at $x = x_f$. Therefore, after the field responses for two uniform sets are summed, the peak value of the resulting main lobe will always be larger than the main lobe peak of each angle set. When $d_{\alpha,1}$ and $d_{\alpha,2}$ are chosen to locate the GLs of two uniform sets in different locations (i.e., $m_1 \lambda/d_{\alpha,1} \neq m_2 \lambda/d_{\alpha,2}$ where m_1 and m_2 are integer numbers), the GLs of the non-uniform angle set will have smaller magnitudes than the main lobe. In addition, even when the GLs of two uniform sets overlap (i.e., $m_1 \lambda/d_{\alpha,1} = m_2 \lambda/d_{\alpha,2}$), they can also be reduced after they are combined if the two coincident GLs have opposite phases. For example, when N_1 and N_2 are both even, the GL at the same location in the beam pattern of the non-uniform set will have a magnitude of:

$$|\psi| = |(-1)^{m_1} N_1 + (-1)^{m_2} N_2|, \tag{6}$$

which can be derived by Equation (3). In this case, the magnitude will be decreased to $|N_1 - N_2|$ if m_1 is even and m_2 is odd, or vice versa.

Figure 3 shows an example of a non-uniform angle set that is obtained by combining two uniform angle sets. The angle distributions (left panels) and synthetic Tx beam patterns (right panels) of the uniform angle set 1 and set 2 are presented in Figure 3a,b, respectively. The number of PWs and angular intervals of set 1 and set 2 are ($N_1 = 6$, $d_{\alpha,1} = 0.069$) and ($N_2 = 6$, $d_{\alpha,2} = 0.046$), respectively. The combined non-uniform angle set and its synthetic beam pattern are shown in Figure 3c. The synthetic beam pattern was obtained by Equation (1), assuming a center frequency of 5.208 MHz and a sound speed of 1540 m/s (i.e., $\lambda = 0.296$ mm).

In Figure 3a,b, the main lobe is located at 0 and the GLs are repeated at a certain interval. The intervals between the GLs of set 1 and set 2 are 4.29 mm and 6.43 mm, respectively, according to Equation (2). The GLs of uniform sets 1 and 2 at different locations are halved in the beam pattern of the non-uniform angle set (see the right panel of Figure 3c). For the chosen parameters ($d_{\alpha,1} = 0.069$, $d_{\alpha,2} = 0.046$), $m_1 \lambda/d_{\alpha,1}$ is equal to $m_2 \lambda/d_{\alpha,2}$ when $m_1 = 3$ and $m_2 = 2$, which means that the third GL of set 1 coincides with the second GL of set 2, as in Figure 3a,b. As the two GL have the same magnitude with the opposite sign according to the Equation (3), they are canceled out in the combined transmit beam pattern (Figure 3c), as is expected from Equation (6). On the other hand, the next coincident GLs at $x = 25.6$ mm, corresponding to $m_1 = 6$ and $m_2 = 4$, have the same sign and thus have the same magnitude as the main lobe peak in the beam pattern of the non-uniform set. However, this theoretical beam pattern is calculated assuming an infinite aperture. Thus, when using a practically used finite-length transducer, these high GLs can also be removed by placing them in a region where fewer or no PWs pass through, which is shown in the Results Section, where the finite length of the array is considered.

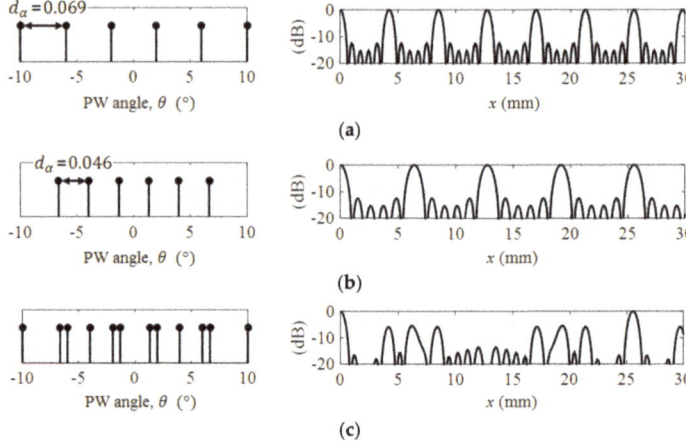

Figure 3. PW angles for synthetic focusing (left panels) and theoretical synthetic beam patterns (right panels) using three angle sets; (**a**) uniform angle set 1: $N_1 = 6$, $d_{\alpha,1} = 0.069$; (**b**) uniform angle set 2: $N_2 = 6$, $d_{\alpha,2} = 0.046$; (**c**) non-uniform angle set (combination of two uniform sets in (**a**,**b**)): $N = N_1 + N_2 = 12$.

3. Results and Discussion

Both of the GL suppression methods were verified through computer experiments by obtaining CW beam patterns and PSF images using a 128-element linear array transducer with a center frequency of 5.208 MHz and a pitch of 0.298 µm (length of the array = 38.1 mm). The CW beam pattern was calculated using MATLAB R2015a (MathWorks, Natick, MA, USA); acoustic field responses of plane waves with different angles, each with a frequency of 5.208 MHz generated from the linear array transducer, were calculated and then compounded with proper delays for a Tx focal point at ($x = 0$, $z = 30$ mm) [1]. The amplitude of each PW was set to 1.0, and the final synthesized beam pattern was normalized by its maximum value and displayed in logarithmic scale.

In the PSF experiments, a single pulse plane wave with a center frequency of 5.208 MHz was transmitted along different directions from the same array transducer. First, RF echo data sets of different plane wave angles reflected from a single point target at ($x = 0$, $z = 30$ mm) were generated using the Field II simulator [15]. Then, MATLAB was used to obtain the PSF images by reconstructing a B-mode image of the point target from the RF data sets using coherent plane wave beamforming [1], demodulation, and logarithmic compression. In the beamforming process, traditional dynamic Rx focusing was performed with an F-number of 1.0, and all the plane waves were coherently synthesized with proper delays for each pixel to obtain two-way (Tx and Rx) dynamic focusing. In the logarithmic compression, the PSF image was normalized by its maximum value and compressed with a dynamic range of 60 dB.

To validate the methods for GL suppression with a uniform d_α, CW Tx beam patterns were obtained, varying the interval, d_α, that is controlled by the number of PW angles within the range of $[-10°, 10°]$. The focus of the beam pattern was at a depth of 30 mm ($z_f = 30$ mm) on the center scanline ($x_f = 0$). Figure 4a presents the magnitude of the first GL relative to that of the main lobe as a function of normalized d_α, $\hat{d}_\alpha$ ($= d_\alpha/(2\lambda/D)$); normalization was performed so that $\hat{d}_\alpha$ equals 1 when Equation (5) is satisfied. Figure 4b illustrates the PW propagating regions; region A is a region where no PWs pass through; region B is a region where more than one PW propagate, and region C is a region where all of the collimated beam areas of the PWs overlap. As mentioned above, the larger $\hat{d}_\alpha$ yields GLs closer to the main lobe. Sections A, B, and C shown in Figure 4a indicate that the first GL is located in regions A, B, and C in Figure 4b, respectively. Figure 4c shows the beam patterns of three values of $\hat{d}_\alpha$ from section A, B, and C (from left to right panels). The gray dash-dot line of each panel indicates the theoretical first GL position ($x_{GL,1} = \lambda/d_\alpha$). From the left to right panels in Figure 4c, one can observe that the magnitude of GL increases as $\hat{d}_\alpha$ increases. Since none of the PWs propagate with the LWF in region A of Figure 4b (i.e., the GL elimination requirement in Equation (4) is satisfied), the GL level is sufficiently suppressed in section A of Figure 4a. In section B of Figure 4a, as $\hat{d}_\alpha$ increases, the first GL moves closer to the focus. As it is closer to the center scanline ($x = 0$) across region B of Figure 4b, the number of overlapped collimated beam areas (i.e., the number of PWs preserving the LWF) increases. Consequently, the GL level increases with $\hat{d}_\alpha$ in section B of Figure 4a. In section C of Figure 4a, the first GL has the same magnitude (0 dB) as the main lobe because the number of PWs passing through each point are the same over all of region C. Note that the GL can be suppressed below −6 dB if $\hat{d}_\alpha < 1$, as in Figure 4a.

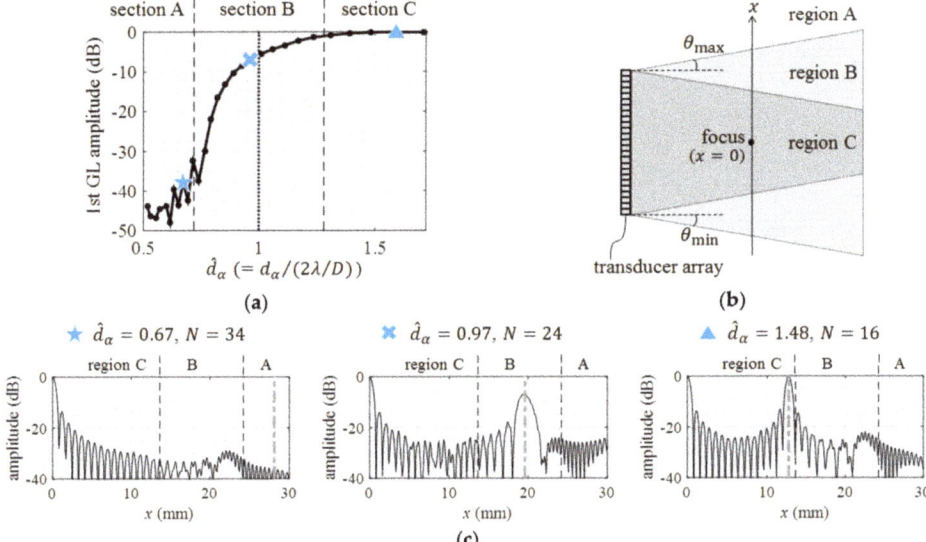

Figure 4. Magnitude of the first GL normalized by that of main lobe when the GL is located in regions A, B, and C. (a) Magnitude of the first GL in a simulated Tx beam pattern versus $\hat{d}_\alpha$. (b) Region where none of (region A), more than one of (region B), and all of (region C) the collimated beam areas of PWs overlap. (c) Simulated beam patterns when using $\hat{d}_\alpha$ (marked with star, cross, and triangle in a (from left to right panels)). The gray dash-dot line indicates the theoretical first GL position, $x_{GL,1}$, of each beam pattern.

The method using a non-uniform angle set was verified by the CW Tx beam pattern and PSF image simulations using the same PW sets used in Figure 3. Figure 5a shows the CW Tx beam patterns using the PW angle sets presented in the left panels of Figure 3a–c, from top to bottom. Since a finite-length transducer is used, regions where not all of the PWs pass through exist. In Figure 5a, region A and B indicate the region where none of the PWs pass and the region where more than one of the PWs propagate, respectively. The GL levels decrease in regions A and B; it is noteworthy that the GL at x = 25.6 mm in Figure 3c is greatly suppressed in the bottom left panel of Figure 5 when considering the finite aperture. The effect of GL reduction using the non-uniform angle set was also assessed with the 2D PSF images shown in Figure 5b. In the top and middle panels of Figure 5b, artifacts due to the first GL are observed, though the GL artifacts were reduced because of the Rx beamforming. The GL artifacts in the middle panel are located further from the point target than those in the top panel as the uniform PW set 2 has a smaller angular interval than that of the PW set 1 ($d_{\alpha,1} = 0.069$ vs. $d_{\alpha,2} = 0.046$).

When the non-uniform angle set is employed (bottom panel in Figure 5b), it is clearly observed that the GLs are successfully suppressed compared to those from the uniform PW sets.

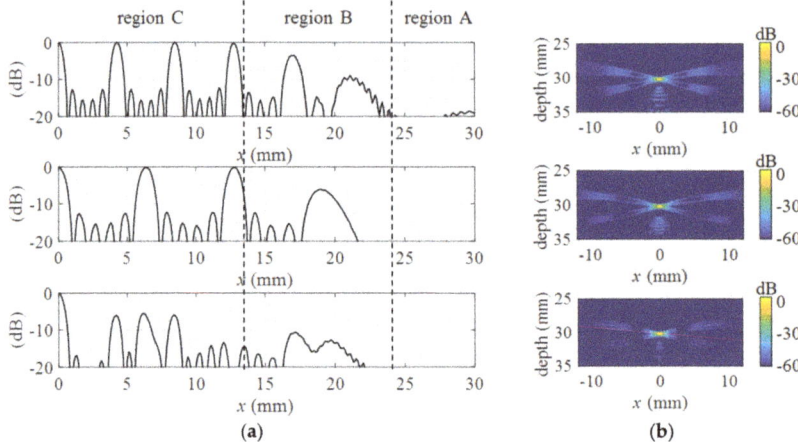

Figure 5. (a) Simulated Tx beam patterns and (b) 2D PSFs of round-trip focusing using the PW angle sets shown in Figure 3a–c (from top to bottom panels).

The proposed non-uniform angle is very easy to design as it is composed of two uniform angle sets with different angle intervals. However, there must be ways to discover other types of non-uniform PW angle distributions or to design an optimal PW angle set for a given imaging specification and the system requirements. One such approach might be to find a specific PW angle distribution that produces GLs where they can be further suppressed by the receive beam pattern, which may also have to be properly (or deliberately) designed. It is also worth noting that a deep neural network can be applied to find a method for the compressed sampling of PW angles [16].

In future studies, we should also consider ways to maintain or improve other important image qualities with a smaller number of PWs for fast imaging. The proposed methods in this paper can be combined with adaptive beamforming and sidelobe reduction techniques [9–14] that have recently been developed to improve the spatial and contrast resolutions of PW imaging. In addition, encoded PWs can be transmitted and compounded for a higher signal-to-noise ratio using Chirp, Barker code, or the Hadamard matrix [5,17,18].

4. Conclusions

In this paper, we describe two conditions for GL formation and present two methods for GL suppression. The first method is proposed for the case in which uniformly distributed angles are used. This method can completely eliminate GL problems, but may have to use a very small angle interval (or, equivalently, too many angles). In such a case, the second method using a non-uniform angle set could be a practical alternative solution for GL suppression. The two methods were verified by CW Tx beam patterns and broad-band 2D PSF images obtained by computer simulations. Future work needs to focus on designing other types of non-uniform PW angle sets to improve the overall image quality in combination with advanced receive beamforming and signal processing techniques.

Author Contributions: T.K.S. devised the algorithm, and S.B. contributed to the design of the experiments. All of the authors wrote and reviewed the manuscript.

Funding: This research was supported by the Next-generation Medical Device Development Program for Newly-Created Market of the National Research Foundation (NRF) funded by the Korean government, MSIP(NRF-2015M3D5A1065997).

Conflicts of Interest: The authors declare no conflict of interest.

References

1. Montaldo, G.; Tanter, M.; Bercoff, J.; Benech, N.; Fink, M. Coherent plane-wave compounding for very high frame rate ultrasonography and transient elastography. *IEEE Trans. Ultrason. Ferroelectr. Freq. Control* **2009**, *56*, 489–506. [CrossRef] [PubMed]
2. Denarie, B.; Tangen, T.A.; Ekroll, I.K.; Rolim, N.; Torp, H.; Bjåstad, T.; Lovstakken, L. Coherent plane wave compounding for very high frame rate ultrasonography of rapidly moving targets. *IEEE Trans. Med. Imaging* **2013**, *32*, 1265–1276. [CrossRef] [PubMed]
3. Lee, J.P.; Song, J.H.; Song, T.-K. Analysis of ultrasound synthetic transmit focusing using plane waves. *J. Acoust. Soc. Korea* **2014**, *33*, 200–209. [CrossRef]
4. ZHao, H.; Song, P.; Urban, M.W.; Greenleaf, J.F.; Chen, S. Shear wave speed measurement using an unfocused ultrasound beam. *Ultrasound Med. Biol.* **2012**, *38*, 1646–1655. [CrossRef] [PubMed]
5. Song, P.; Urban, M.W.; Manduca, A.; Greenleaf, J.F.; Chen, S. Coded excitation plane wave imaging for shear wave motion detection. *IEEE Trans. Ultrason. Ferroelectr. Freq. Control* **2015**, *62*, 1356–1372. [CrossRef] [PubMed]
6. Udesen, J.; Gran, F.; Hansen, K.L.; Jensen, J.A.; Thomsen, C.; Nielsen, M.B. High frame-rate blood vector velocity imaging using plane waves: Simulations and preliminary experiments. *IEEE Trans. Ultrason. Ferroelectr. Freq. Control* **2008**, *55*, 1729–1743. [CrossRef] [PubMed]
7. Bercoff, J.; Montaldo, G.; Loupas, T.; Savery, D.; Mézière, F.; Fink, M.; Tanter, M. Ultrafast compound Doppler imaging: Providing full blood flow characterization. *IEEE Trans. Ultrason. Ferroelectr. Freq. Control* **2011**, *58*, 134–147. [CrossRef] [PubMed]
8. Tong, L.; Gao, H.; Choi, H.F.; D'hooge, J. Comparison of conventional parallel beamforming with plane wave and diverging wave imaging for cardiac applications: A simulation study. *IEEE Trans. Ultrason. Ferroelectr. Freq. Control* **2012**, *59*, 1654–1663. [CrossRef] [PubMed]
9. Austeng, A.; Nilsen, C.I.C.; Jensen, A.C.; Näsholm, S.P.; Holm, S. Coherent plane-wave compounding and minimum variance beamforming. In Proceedings of the IEEE International Ultrasonics Symposium, Orlando, FL, USA, 18–21 October 2011; pp. 2448–2451.
10. Zhao, J.; Wang, Y.; Zeng, X.; Yu, J.; Yiu, B.Y.S.; Yu, A.C.H. Plane wave compounding based on a joint transmitting-receiving adaptive beamformer. *IEEE Trans. Ultrason. Ferroelectr. Freq. Control* **2015**, *62*, 1440–1452. [CrossRef] [PubMed]
11. Zimbico, A.J.; Granado, D.W.; Schneider, F.K.; Maia, J.M.; Assef, A.A.; Schiefler, N.; Costa, E.T. Eigenspace generalized sidelobe canceller combined with SNR dependent coherence factor for plane wave imaging. *Biomed. Eng. Online* **2018**, 1–23. [CrossRef] [PubMed]
12. Guo, W.; Wang, Y.; Yu, J. A sibelobe suppressing beamformer for coherent plane wave compounding. *Appl. Sci.* **2016**, *6*, 359. [CrossRef]
13. Wang, Y.; Zheng, C.; Peng, H.; Zhang, C. Coherent plane-wave compounding based on normalized autocorrelation factor. *IEEE Access* **2018**, *6*, 36927–36938. [CrossRef]
14. Guo, W.; Wang, Y.; Wu, G.; Yu, J. Sidelobe reduction for plane wave compounding with a limited frame number. *Biomed. Eng. Online* **2018**, *17*, 94. [CrossRef] [PubMed]
15. Jensen, J.A. Field: A program for simulating ultrasound systems. In Proceedings of the 10th Nordic-Baltic Conference on Biomedical Imaging, Lyngby, Denmark, 9–13 June 1996; Volume 34, pp. 351–353.
16. Gasse, M.; Millioz, F.; Roux, E.; Garcia, D.; Liebgott, H.; Friboulet, D. High-Quality Plane Wave Compounding Using Convolutional Neural Networks. *IEEE Trans. Ultrason. Ferroelectr. Freq. Control* **2017**, *64*, 1637–1639. [CrossRef] [PubMed]
17. Zhao, F.; Tong, L.; He, Q.; Luo, J. Coded excitation for diverging wave cardiac imaging: A feasibility study. *Phys. Med. Biol.* **2017**, *62*, 1565–1584. [CrossRef] [PubMed]
18. Tiran, E.; Deffieux, T.; Correia, M.; Maresca, D.; Osmanski, B.-F.; Sieu, L.-A.; Bergel, A.; Cohen, I.; Pernot, M.; Tanter, M. Multiplane wave imaging increases signal-to-noise ratio in ultrafast ultrasound imaging. *Phys. Med. Biol.* **2015**, *60*, 8549–8566. [CrossRef] [PubMed]

© 2018 by the authors. Licensee MDPI, Basel, Switzerland. This article is an open access article distributed under the terms and conditions of the Creative Commons Attribution (CC BY) license (http://creativecommons.org/licenses/by/4.0/).

Article

Coded Excitation for Crosstalk Suppression in Multi-line Transmit Beamforming: Simulation Study and Experimental Validation

Ling Tong [1], Qiong He [1], Alejandra Ortega [2], Alessandro Ramalli [2,3], Piero Tortoli [3], Jianwen Luo [1,*] and Jan D'hooge [2,*]

1. Center for Bio-Medical Imaging Research, Dept. of Biomedical Engineering, School of Medicine, Tsinghua University, Beijing 10084, China; tonglingpku@gmail.com (L.T.); cmuheqiong@163.com (Q.H.)
2. Lab. On Cardiovascular Imaging and Dynamics, Dept. of Cardiovascular Sciences, KU Leuven, 3000 Leuven, Belgium; alejao16@gmail.com (A.O.); alessandro.ramalli@unifi.it (A.R.)
3. Dept. of Information Engineering, University of Florence, 50139 Firenze, Italy; piero.tortoli@unifi.it
* Correspondence: luo_jianwen@tsinghua.edu.cn (J.L.); jan.dhooge@uzleuven.be (J.D.); Tel.: +86-10-6278-0650 (J.L.); +32-16-3-49012 (J.D.)

Received: 29 August 2018; Accepted: 26 January 2019; Published: 31 January 2019

Abstract: (1) Background: Multi-line transmit (MLT) beamforming has been proposed for fast cardiac ultrasound imaging. While crosstalk between MLT beams could induce artifacts, a Tukey ($\alpha = 0.5$)-Tukey ($\alpha = 0.5$) transmit-receive (TT-) apodization can largely—but not completely—suppress this crosstalk. Coded excitation has been proposed for crosstalk suppression, but only for synthetic aperture imaging and multi-focal imaging on linear/convex arrays. The aim of this study was to investigate its (added) value to suppress crosstalk among simultaneously transmitted multi-directional focused beams on a phased array; (2) Methods: One set of two orthogonal Golay codes, as well as one set of two orthogonal chirps, were applied on a two, four, and 6MTL imaging schemes individually. These coded schemes were investigated without and with TT-apodization by both simulation and experiments; and (3) Results: For a 2MLT scheme, without apodization the crosstalk was removed completely using Golay codes, whereas it was only slightly suppressed by chirps. For coded 4MLT and 6MLT schemes, without apodization crosstalk appeared as that of non-apodized 2MLT and 3MLT schemes. TT-apodization was required to suppress the remaining crosstalk. Furthermore, the coded MLT schemes showed better SNR and penetration compared to that of the non-coded ones. (4) Conclusions: The added value of orthogonal coded excitation on MLT crosstalk suppression remains limited, although it could maintain a better SNR.

Keywords: multi-line transmit; crosstalk artifacts; coded excitation; cardiac imaging

1. Introduction

High frame rate imaging has recently gained increased attention in the field of echocardiography given its potential to reveal new areas of myocardial mechanics and blood flow analysis [1]. Among high frame rate imaging approaches, plane wave or diverging wave imaging are popular research topics due to their capacity to produce very high frame rates by scanning a given field-of-view (i.e., a 90-degree sector) with only a few transmissions [2–4]. However, the signal-to-noise ratio (SNR), spatial and contrast resolution of the resulting images are degraded due to the lack of focusing. To compensate this limitation, spatial coherent compounding is generally required in which the same region is interrogated several times from different directions and the final image is an average of all acquisitions [5]. As a drawback, the effective gain in frame rate drops by a factor equal to the number of the compounded images. Moreover, motion artifacts can often occur during compound.

Additionally, plane waves or diverging waves spread energy over a large area leading to small acoustic pressure amplitudes. Hence, it is technically challenging to make harmonic imaging.

As an alternative, multi-line transmit beamforming (MLT) has also been proposed [6,7]. In this approach, multiple focused beams are simultaneously transmitted into different directions leading to a gain in frame rate equal to the number of MLT beams. Typically, to simultaneously obtain multiple focused beams, the pulses that would be applied to individual elements to sequentially generate focused beams at different directions during several transmit events can be literally superimposed and be applied to those elements during a single transmit event. As an example, Figure 1 shows the pulses that would be used to generate four ultrasound beams either sequentially (Figure 1a–d) or simultaneously (i.e., 4-MLT) (Figure 1e). Figure 1f presents the transmit beam pattern corresponding to a case of 4MLT (Figure 1e). In contrast to plane wave/diverging wave imaging, MLT utilizes focused beams. This implies that the resulting SNR and spatial resolution [8,9] can be preserved as well as the possibility of a second harmonic imaging [10]. Despite these advantages, MLT beams may potentially introduce crosstalk that would appear as ghost-like artifacts on the images (Figure 2a). Intrinsically, crosstalk artifacts are the results of the interference between MLT beams in different directions. It has been demonstrated that such crosstalk artifacts can be largely suppressed by using a Tukey ($\alpha = 0.5$)-Tukey ($\alpha = 0.5$) (TT) transmit and receive apodization scheme; so that the resulting MLT images look competitive to the conventional single line transmit beamforming (SLT) (Figure 2b) [8,9,11]. Nonetheless, despite these promising results, residual crosstalk artifacts can sometimes be detected [9]. Recently, alternative approaches, such as minimum variance (MV) receive beamforming [12], low complexity adaptive (LCA) receive apodization [13], and filtered delay multiply and sum (F-DMAS) [14] have been proposed to reduce receive crosstalk. Indeed, MV adaptive beamforming could not obtain the same crosstalk reduction as simply using a Tukey apodization when received, though a better spatial resolution could be obtained [12]. Similar, the LCA adaptive apodization method using a modified predefined apodization bank could have slightly better crosstalk reduction while improving the contrast and spatial resolution, but artifacts remained visible when hyperechoic structures were presented (for instance the pericardium) [13]. The F-DMAS method could provide a better receive crosstalk suppression but the contrast-to-noise ratio would be degraded [14]. Nonetheless, more attempts for better reduction of crosstalk remain desired.

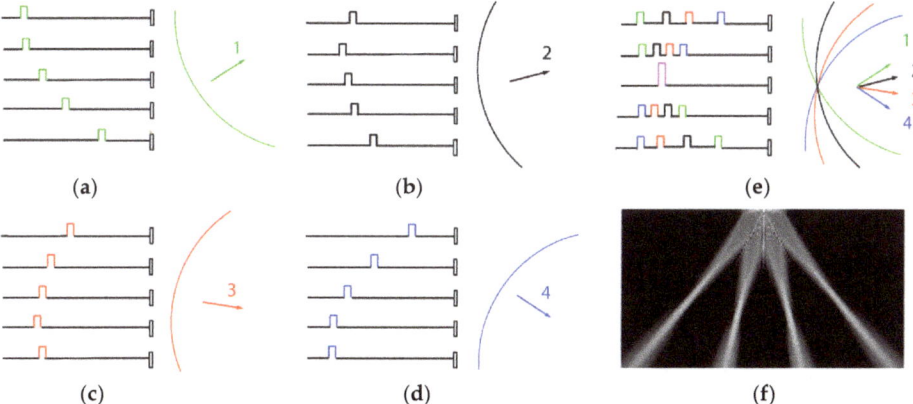

Figure 1. Pulses to be applied on individual elements of a phased array transducer in order to generate four focused transmit beams: (**a**) Transmit direction 1, (**b**) transmit direction 2, (**c**) transmit direction 3, (**d**) transmit direction 4, consecutively, or (**e**) simultaneously; and (**f**) A beam pattern of 4 simultaneously transmit beams.

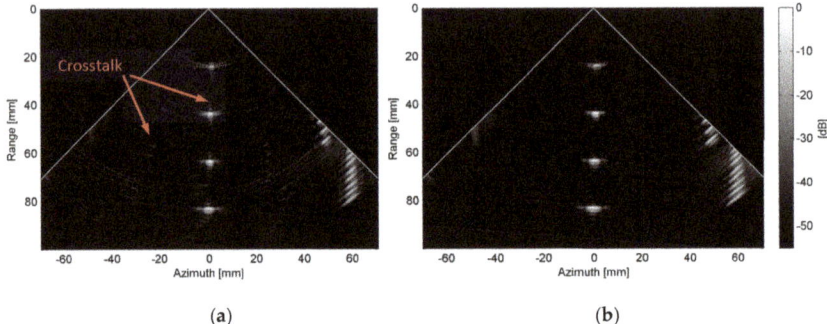

Figure 2. Reconstructed images of a wire phantom in water acquired using a 4-MLT (Multi-line transmit) imaging scheme (**a**) without apodization, the ghost-like crosstalk artifacts are presented; and (**b**) with a Tukey ($\alpha = 0.5$)-Tukey ($\alpha = 0.5$) apodization on transmit and receive, the crosstalk artifacts are suppressed significantly.

Coded excitation has been shown to improve the penetration, signal-to-noise ratio, as well as the frame rate for ultrasound imaging. In terms of increasing frame rate, it has been proposed to suppress crosstalk in many applications [15–21] by using simultaneous transmission, of two or more orthogonal transmit pulses whose mutual cross-correlation is very small. When simultaneously transmitting two or more such mutually orthogonal codes, the received signals contain information from all codes. This information can then be separated through decoding process. Common orthogonal codes are the Golay or chirp codes. Golay codes possess a good orthogonal property, but require the transmission of complementary code pairs to constrain the range lobes, which in turn halve the effective frame rate. On the other hand, orthogonal chirp pairs can be obtained by sweeping a certain frequency band in opposite directions or by sweeping separate frequency (sub)bands. For the former case, the cross-correlation of two orthogonal chirps can be relatively high, whereas the imaging performance of the latter is typically limited by the narrow bandwidth of ultrasound transducers. However, for chirps, the transmission of complementary codes is not necessary, and so the frame rate would be compromised. Both orthogonal Golay and chirp codes have been proposed for crosstalk reduction in increasing frame rate for synthetic aperture imaging with multiple transmit positions [15,17] and in maintaining frame rate for multi-zone focusing by simultaneous transmissions for linear [16–19,22] and convex array imaging [20,21]. However, towards phased array based high frame rate cardiac imaging, it has not been revealed yet that the feasibility of using these orthogonal codes to (further) reduce crosstalk among focused MLT beams. Hence, the aim of this study was to test this feasibility in both simulation and experimental setups.

2. Materials and Methods

In this study, a typical 1-D cardiac phased array probe (PA230, Esaote SpA, Florence, Italy) with 128 elements, a central frequency of 2.0 MHz and a bandwidth of 50% was used both for simulation and experiments. The array measured 21.6 mm in width and 13 mm in height with a pitch of 170 μm. To accommodate our experimental system, which possesses 64 independent channels, only the odd elements were pinned, resulting in an equivalent 64-element phased array with an effective pitch of 340 μm. This choice was made so as to exploit the maximum aperture of the probe (22 mm) with the 64 available channels on the system; the selection of 64 consecutive elements would have reduced the aperture size to 11 mm, thus limiting the lateral resolution that is already poor in cardiac phased array imaging. Moreover, for a central frequency of 2.0 MHz, the corresponding central wavelength of the probe is 770 μm at a speed of sound of 1540 cm/s. Thus, the effective pitch of 340 μm is about half of the central wavelength. This limits the possibility to produce grating lobes when steering the beams out at 45 degrees, i.e., the typical maximum steering angle in cardiac imaging.

2.1. Orthogonal Coded Excitations

On this probe, two types of orthogonal coded excitations, i.e., one set of orthogonal Golay codes and one set of orthogonal linear frequency (FM) modulated chirps were tested since they have already been proposed for other purposes in ultrasound imaging [12–19] and are relatively simple to implement. In particular, the following Golay codes were used to obtain good orthogonal property without largely elongating the excitation duration:

G1: [1, 1, 1, −1]; G1c: [1, 1, −1, 1]; G2: [1, −1, 1, 1]; and G2c: [1, −1, −1, −1].

where G1 is complementary to G1c, G2 is complementary to G2c, the pair of G1 and G1c are orthogonal to the pair of G2 and G2c. The excitations were obtained by convolving every Golay codes with a burst of 1.45 cycles square wave at the central frequency of the transducer, respectively, that resulted in a duration of 2.91 µs.

A linear FM chirp coded excitation can be defined as:

$$c(t) = a(t) \cdot exp\left[j2\pi\left(f_0 t + \frac{B}{2T}t^2\right)\right], -\frac{T}{2} \leq t \leq \frac{T}{2}, \quad (1)$$

where $a(t)$ is the tapering function, f_0 is the central frequency, B is the bandwidth of the chirp signal, and T is the signal duration. In our case to have better orthogonality, $a(t)$ was a Tukey window ($\alpha = 0.2$), T was 10 µs, and B was 3.8 MHz centered around 2 MHz. Two orthogonal chirps, c_{up} and c_{down}, were obtained by sweeping B with the opposite directions. The different excitation signals are sketched in Figure 3. Note that the large sweeping bandwidth of the chips codes were chosen to minimize the Fresnel rippes around the pass-band of the probe for a better response, as indicated in Figure 4. Moreover, to decode, matched filters were used for the Golay codes, whereas Chebyshev-apodized mis-matched filters were adopted for the chirps.

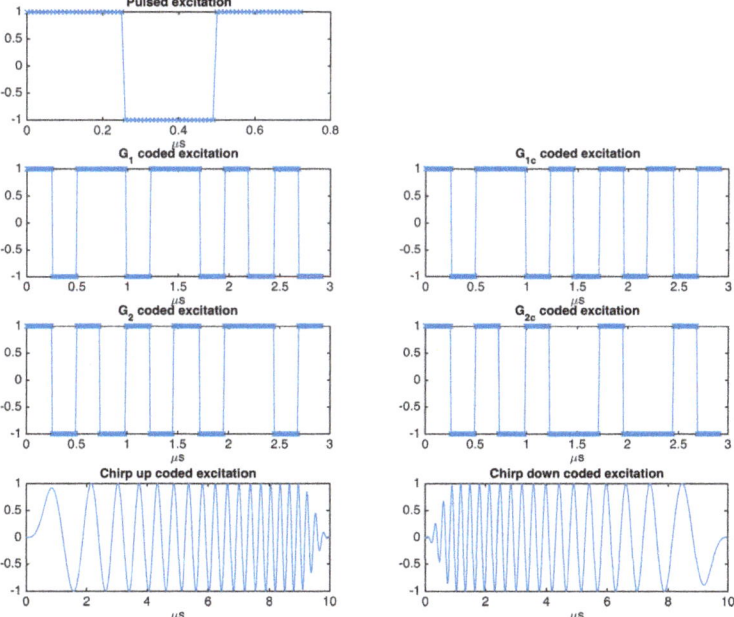

Figure 3. Plots of different excitation signals.

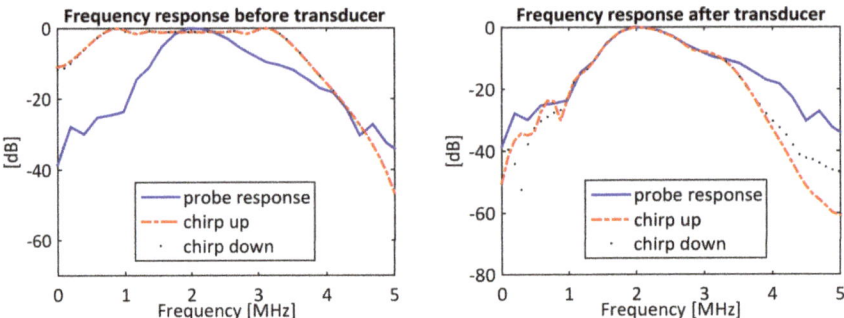

Figure 4. Frequency response of the two chirp codes.

2.2. Coded Excitation Based Multi-Line Transmit (MLT) Imaging Schemes

Using the two sets of orthogonal codes, a 2, 4, and 6-MLT imaging scheme were set up. The MLT transmits were obtained by superimposing the transmit pulse patterns that would be applied to the probe elements to generate subsequent transmit beams [8]. The imaging sector was set to 90° and covered by 180 image lines. For a given MLT imaging scheme, an orthogonal code pair was applied on neighboring MLT beams in an interleaved manner, as illustrated in Figure 5, i.e., neighboring sub-sectors were scanned using MLT beams with orthogonal codes. In this way, signals generated from neighboring MLT beams were expected to be better differentiated through decoding in reception, which was performed before beamforming.

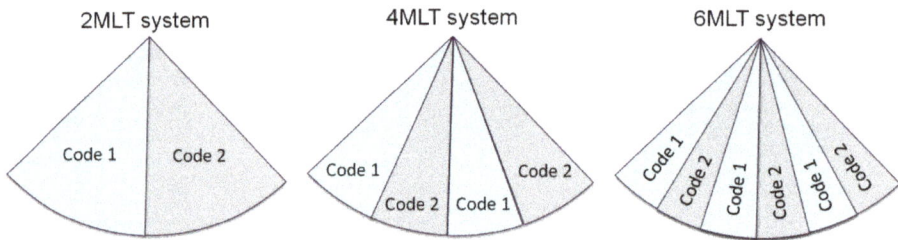

Figure 5. Illustration of the interleaving arrangement of two orthogonal codes on a given MLT imaging scheme. Code 1 and code 2 are mutual orthogonal codes.

2.3. Simulations

The performance of coded excitation imaging scheme was first investigated in silico through the simulation software Field II [23,24] based on the phased array described at the beginning of this section. To benchmark, an SLT imaging scheme, 2, 4, and 6-MLT schemes using a burst excitation (1.45 cycles square wave at 2 MHz) were included. For all imaging schemes, the transmit focal point was maintained at a depth of 70 mm and a dynamic apodization with f# of two was adopted in reception. As previously proposed, a Tukey (α = 0.5) -Tukey (α = 0.5) (TT) apodization scheme on transmit and receive were also tested on different MLT schemes (coded and non-coded) for crosstalk suppression [8,9].

To appreciate the spatial resolution as well as the crosstalk artifacts, the point-spread-functions (PSFs) of different imaging schemes were simulated. In particular, four point scatterers were positioned at depths from 30 mm to 90 mm with an equal axial interval of 20 mm. In the azimuth direction, they were positioned at 10 mm offset of the transducer symmetry axis to better appreciate the potential crosstalk. A given PSF profile was evaluated by the −6 dB beam width in both radial and lateral directions. The crosstalk was quantified by calculating the intensity difference between the PSF of a given MLT scheme and that of the SLT imaging scheme ($xtalk_{psf}$). For details on the definitions of these parameters, please refer to Formulas (3), (5), and (7) in Reference [9].

2.4. Experimental Validation

To experimentally investigate the performance of the coded MLT schemes, the imaging schemes defined in Section 2.3 were implemented on the ultrasound advanced open platform (ULA-OP) [25,26] equipped with the above mentioned phased array transducer. The pulse repetition frequency (PRF) was 4750 Hz and the prebeamformed radio-frequency (RF) channel data were acquired at a sampling frequency of 50 MHz using the ULA-OP acquisition board [27] and beamformed offline through MATLAB R2015a (The MathWorks, Natick, MA, USA). The beamforming parameters were the same as those applied in the simulations.

Using these setups, a general-purpose tissue mimicking phantom (CIRS Model 040GSE, Norfolk, VA, USA) and the left ventricle of a healthy volunteer from a parasternal long axis view were imaged. To quantify the image quality, the contrast-to-noise (CNR), the contrast ratio (CR) of the cystic regions, and the signal-to-noise ratio (SNR) were calculated for the static phantom images:

$$CNR = \frac{\mu_b - \mu_c}{\sqrt{\sigma^2_b + \sigma^2_c}}, \quad (2)$$

$$CR = 20 log_{10} \frac{\mu_b}{\mu_c} \quad (3)$$

$$SNR = 10 log_{10} \frac{\mu_b}{\mu_n} \quad (4)$$

where μ_b and μ_c, μ_n, are the mean gray-level in the background, the cystic regions and in the noise region, respectively, and σ_b and σ_c are their corresponding standard deviations. Finally, the quality of the in-vivo images was visually examined.

3. Results

The simulated PSFs of different MLT schemes without and with coded excitation are presented in Figures 6–8. Without apodization on transmit and receive, no crosstalk was seen in the image of the Golay-coded 2-MLT scheme (Figure 6c) within a 55 dB-dynamic range, whereas a slight crosstalk suppression was observed in the image of the chirp-coded 2-MLT scheme (Figure 6d, highest crosstalk at about −46 dB) compared to that of the non-coded 2-MLT scheme (Figure 6b, highest crosstalk at about −40 dB). For the non-apodized coded 4-MLT and 6-MLT schemes, obvious crosstalk artifacts was observed (Figure 7c–f), though less artifacts were found on the Golay-coded images than on the chirp-coded images in particular in the near field. With TT apodization, most of the crosstalk was suppressed within the 55 dB-dynamic range (Figure 8), despite the fact that the image of the point scatterer at the depth of 30 mm was severely distorted for the TT-apodized chirp-coded 6MLT scheme. Quantitative crosstalk evaluation is shown in Figure 9. It was found that, without apodization and for the same number of MLT beams, Golay-coded MLT schemes presented the lowest crosstalk level. In particular, zero level crosstalk was detected for Golay-coded 2-MLT schemes. Moreover, with apodization, the crosstalk level of both Golay and chirp-coded 4MLT and 6MLT schemes was significantly reduced compared to that of the non-apodized ones. In general, despite of the apodized chirp-coded 6MLT schemes, all apodized coded MLT schemes showed lower crosstalk compared to the corresponding non-apodized non-coded MLT schemes.

In addition, for the mean radial beam widths, all imaging schemes showed similar values (around 1mm), although Golay-coded schemes had slightly larger values (about 1.07 mm) (Figure 10, top plot). With respect to lateral beam widths, the non-apodized coded schemes showed similar widths, which was about 2.30 mm, whereas the values of apodized schemes were around 2.89 mm (Figure 10, bottom plot).

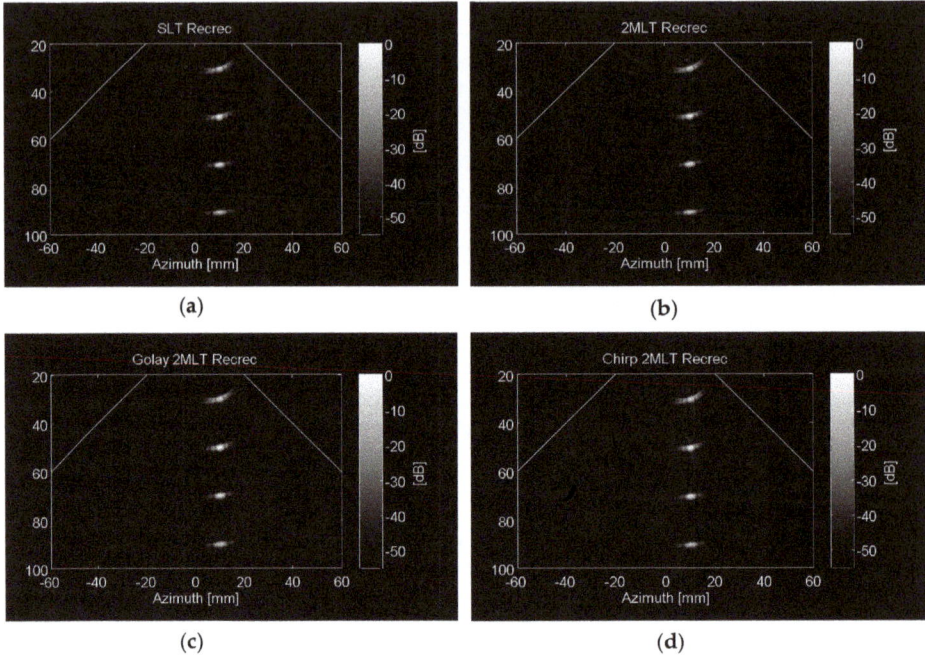

Figure 6. Point-spread-function (PSF) of different MLT schemes without apodization on transmit and receive. (**a**) SLT (single line transmit i.e., 1MLT) scheme with non-coded excitation as the reference image, (**b**) 2MLT scheme using non-coded excitation with crosstalk artifacts, (**c**) 2MLT scheme using Golay coded excitation without crosstalk artifacts, and (**d**) 2MLT scheme using chirp coded excitation with slight crosstalk artifacts. A dynamic range of 55 dB was used.

Based on the simulation results, TT-apodized 4MLT and 6MLT schemes were tested in in-vitro experiments. Results are presented in Figures 11 and 12 where in Figure 12 the cystic regions were zoomed for better visualization. The images of the tissue phantom presented in Figure 11 showed that the crosstalk artifacts were hardly observed for all MLT schemes. However, the coded images showed better contrast for both the middle and bottom cystic regions (Figure 12). Moreover, coded images presented less electronic background noise compared to the non-coded ones (Figure 11, indicated by the arrows). Quantitatively, seven regions of interest (ROI) were defined on those images to compute the SNR, CNR, and CR. As indicated in Figure 11a, three ROIs were defined in the three cystic regions by adjusting the radius of the cysts, whereas three ROIs were accordingly defined in the hyperechoic background next to individual cysts. The CNR and CR of each cystic region were thus computed using Equations (2) and (3). Moreover, the last ROI was defined in the noise region (Figure 11, N_{ROI}), where the SNR was computed based on signal in N_{ROI} and B_{ROI} using Equation (4). The results are presented in Table 1. It can be seen that in terms of CNR, all MLT imaging schemes showed similar values for the top and middle cysts, whereas for the bottom cyst, the coded 4-MLT and 6-MLT schemes showed higher values than that of the non-coded 4-MLT and 6-MLT schemes, respectively. In terms of CR, except for the top cyst, coded schemes showed about 2.1–5.5 dB higher CR compared to that of the non-coded ones; the chirp-coded MLT schemes presented the highest CR values. Similarly, with respect to SNR, the coded schemes had about 2.6–5.4 dB higher SNR than that of the non-coded ones; still, the Golay-coded performed the best and presented the highest SNR.

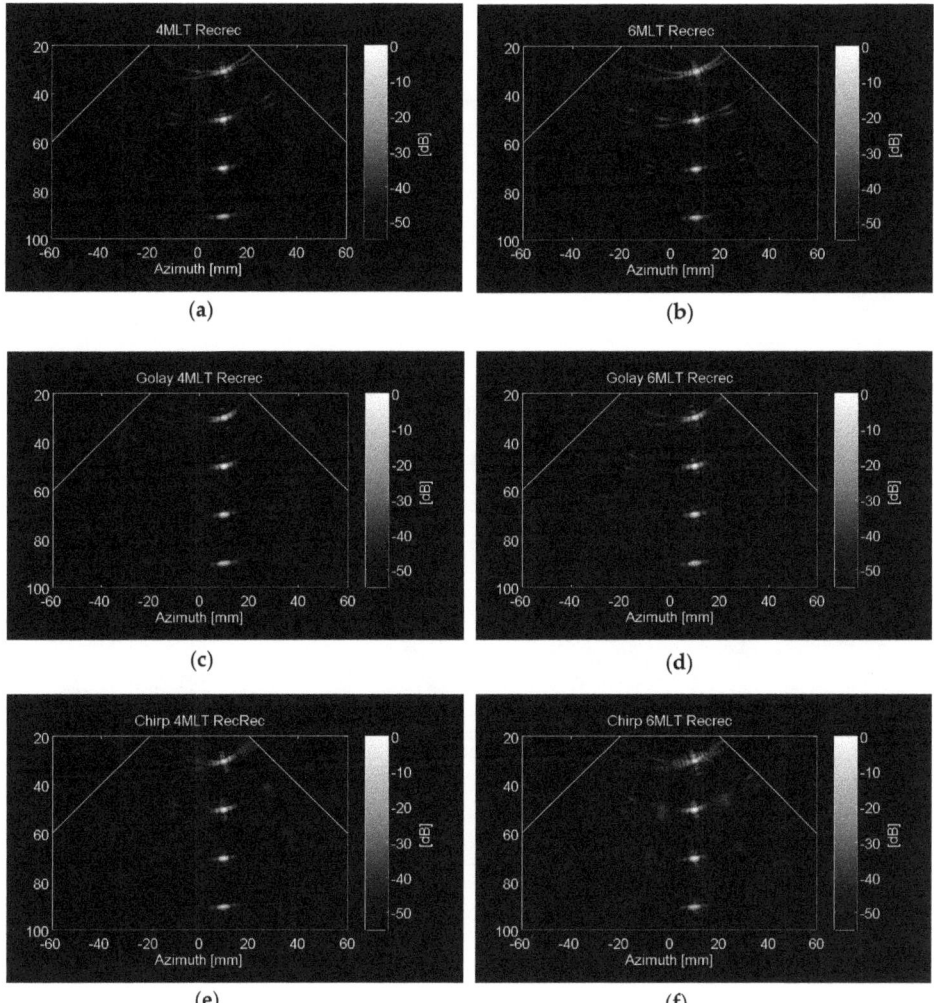

Figure 7. Point-spread-function (PSF) of 4MLT and 6MLT schemes using (**a**,**b**) non-coded excitation; (**c**,**d**) Golay coded excitations; and (**e**,**f**) chirp coded excitations. No apodization was applied on transmit and receive. Crosstalk artifacts appeared and became more severe as increasing the MLT beams.

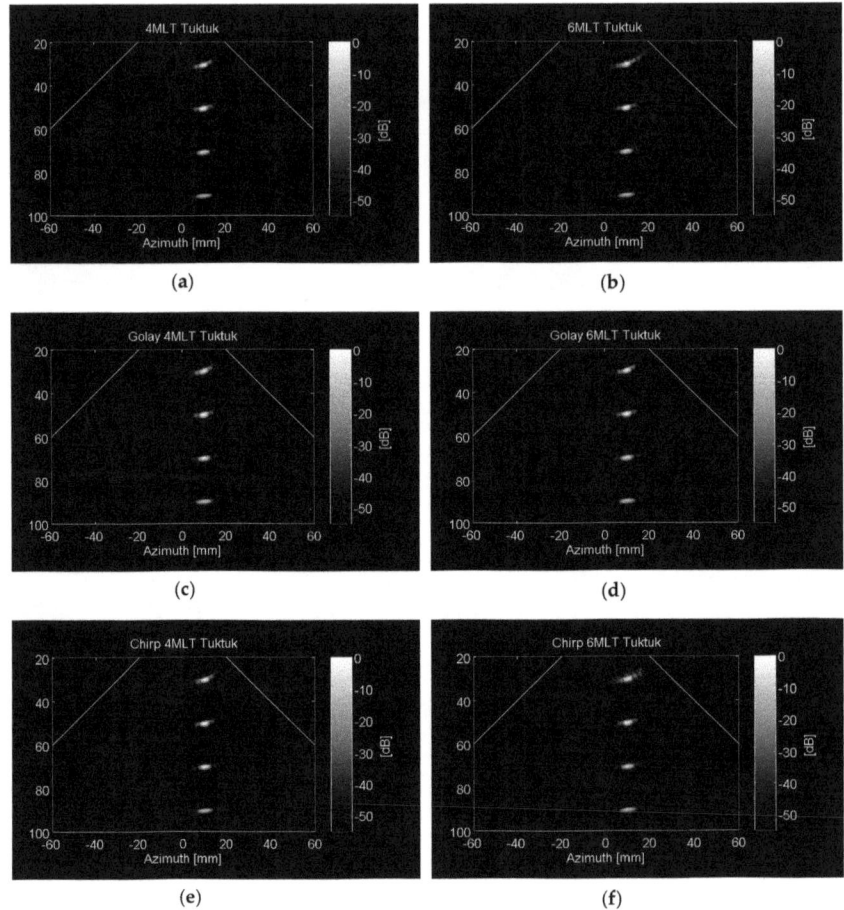

Figure 8. Point-spread-function (PSF) of 4MLT and 6MLT schemes using (**a,b**) non-coded excitation; (**c,d**) Golay coded excitations; and (**e,f**) chirp coded excitations. Tukey ($\alpha = 0.5$)-Tukey ($\alpha = 0.5$) apodization scheme was applied on transmit and receive that the crosstalk artifacts were largely suppressed.

Table 1. Quantifications of different Tukey-Tukey apodized multi-line transimt (MLT) schemes.

Imaging Schemes	Non-Coded 4-MLT	Non-Coded 6-MLT	Golay-Coded 4-MLT	Golay-Coded 6-MLT	Chirp-Coded 4-MLT	Chirp-Coded 6-MLT
CNR_{top}	1.28	1.31	1.18	1.15	1.24	1.28
CNR_{middle}	1.53	1.40	1.49	1.50	1.56	1.60
CNR_{bottom}	1.82	1.01	2.05	1.31	2.25	1.51
CR_{top} (dB)	9.54	10.75	8.11	9.40	9.93	10.50
CR_{middle} (dB)	17.26	14.49	20.40	17.76	19.41	20.01
CR_{bottom} (dB)	11.55	6.50	14.52	8.12	16.03	9.99
SNR (dB)	12.71	10.66	18.15	16.17	15.94	13.31
Frame rate (Hz)	105.56	158.33	52.78	79.17	105.56	158.33

Finally, in-vivo images of a healthy volunteer were acquired with the three TT-apodized 4MLT schemes (Figure 13). From the cineloops (Supplementary Materials: Video S1: Figure13_a; Video S2: Figure13_b; Video S3: Figure13_c), the coded images seemed to have better contrast between the myocardial and blood pool and no obvious crosstalk was observed.

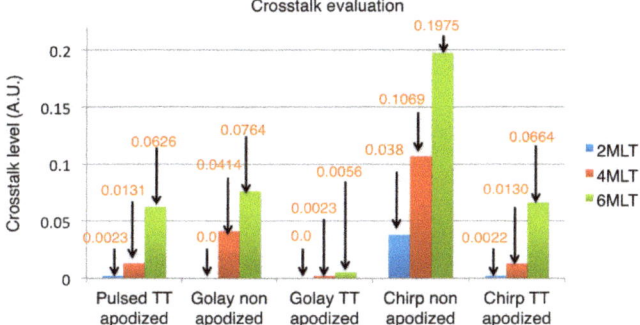

Figure 9. Crosstalk evaluation of different MLT schemes. TT: Tukey ($\alpha = 0.5$)-Tukey ($\alpha = 0.5$) apodization scheme.

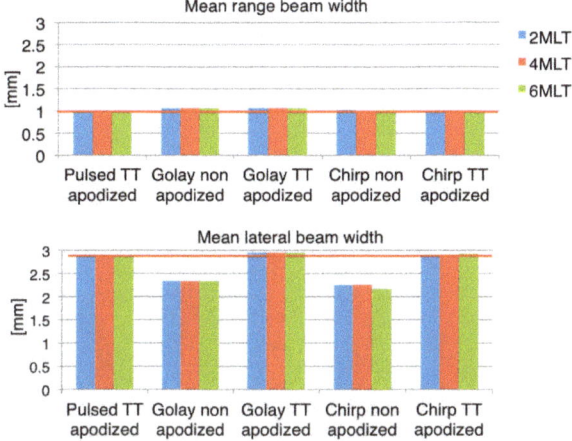

Figure 10. Beam widths in radial (upper panel) and lateral (lower panel) directions estimated from the point-spread-function (PSF). The red line indicates the value of a non-coded SLT scheme using TT apodization. TT: Tukey ($\alpha = 0.5$)-Tukey ($\alpha = 0.5$) apodization scheme.

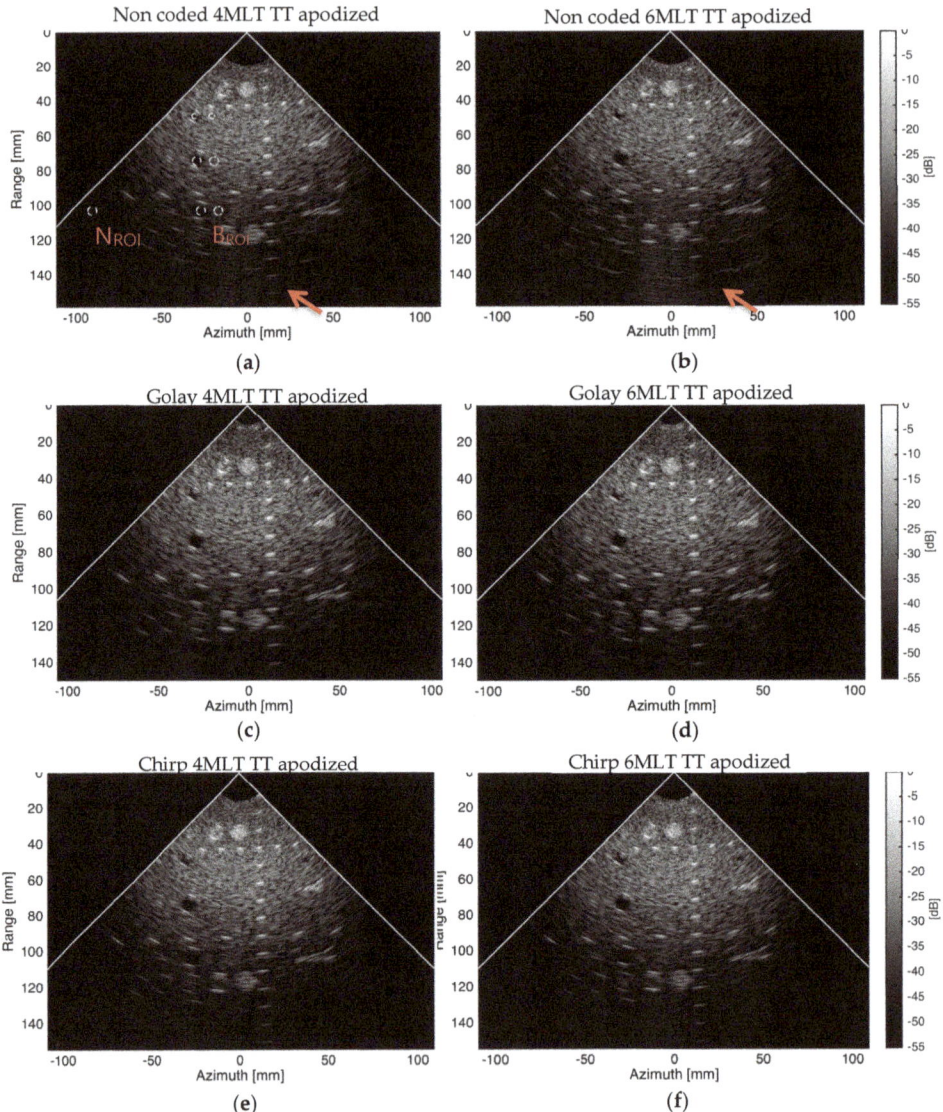

Figure 11. Tissue phantom images of the 4MLT and 6MLT schemes using (**a**,**b**) non-coded; (**c**,**d**) Golay-coded excitation; or (**e**,**f**) chirp-coded excitation. TT apodization was applied on both transmit and receive. TT: Tukey ($\alpha = 0.5$)-Tukey ($\alpha = 0.5$) apodization scheme. A dynamic range of 55 dB was used. The arrows in panel (**a**) and (**b**) point the areas in which the non-coded MLT schemes started to lose signal-to-noise ratio (SNR).

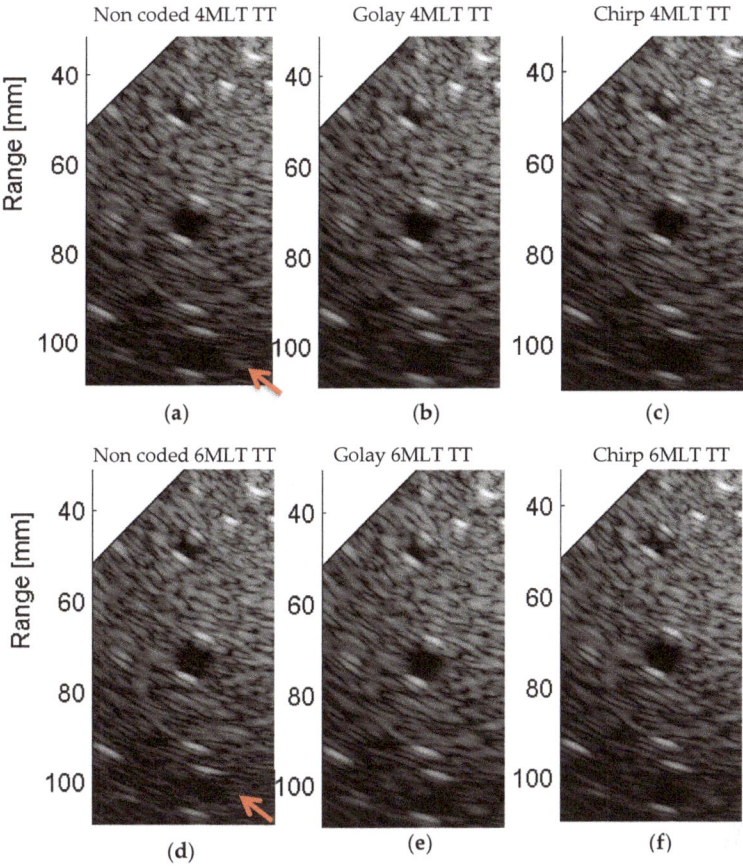

Figure 12. Zoomed in tissue phantom images of the 4MLT and 6MLT schemes using (**a**,**d**) non-coded; (**b**,**e**) Golay-coded excitation; or (**c**,**f**) chirp-coded excitation. TT apodization was applied on both transmit and receive. TT: Tukey ($\alpha = 0.5$)-Tukey ($\alpha = 0.5$) apodization scheme. A dynamic range of 55 dB was used. The arrows in panel (**a**) and (**d**) point the areas in which the non-coded MLT schemes started to lose SNR.

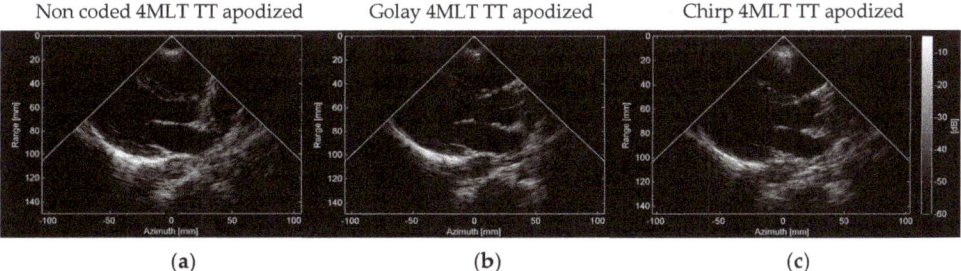

Figure 13. In-vivo examples of 4MLT schemes using (**a**) non-coded excitation, (**b**) Golay coded excitation, and (**c**) chirp coded excitation with TT apodization. TT: Tukey ($\alpha = 0.5$)-Tukey ($\alpha = 0.5$) apodization scheme. Video clips available online. A dynamic range of 55 dB was used (video clips available as Supplementary Materials).

4. Discussion

In this study, we investigated the impact of two types of orthogonal codes on phased array based MLT imaging schemes in order to see whether they could be of (added) value to suppress the associated crosstalk artifacts. The performance of these coded imaging schemes was investigated without and with TT apodization and benchmarked against non-coded MLT schemes with TT apodization in both simulation and experiments. The results show that without apodization, for a 2MLT scheme, the crosstalk was completely removed using orthogonal Golay codes with a 55 dB dynamic range, whereas it was slightly suppressed with the orthogonal chirps (Figures 6 and 9). Indeed, this implied extremely low cross-correlation between the used orthogonal Golay pair. Compared to a non-coded non-apodized four or 6-MLT scheme, a coded non-apodized four or 6-MLT scheme showed less crosstalk artifacts and the appearance of its artifacts largely replicated that of a non-coded non-apodized two or 3-MLT scheme, respectively (Figure 7). This was to be expected given that only two sets of orthogonal codes were applied on neighboring MLT beams in an interleaved manner. Hence, to suppress the remaining crosstalk, TT apodization was applied (Figure 8). While almost no crosstalk artifacts could have been seen on the TT-apodized coded images (Figure 8), crosstalk quantification showed that the chirp-coded schemes had slightly higher crosstalk level than that of the Golay-coded ones (Figure 9). This was in agreement with the fact that the cross-correlation of the two "orthogonal" chirps remained high due to the overlapping sweeping frequency bands. The information in two different directions can thus not be completely separated. Moreover, in the near field, the PSF of the chirp-coded 6MLT scheme distorted [e.g., Figure 8f the scatterer at a depth of 30 mm]. This might be due to the fact that the inter-beam space was very small for the 6MLT scheme when it was close to the probe, since the duration of the chirps were relatively long, the level of interference between simultaneously transmitted waveforms could be too high to be well separated in the near field. Nonetheless, with respect to crosstalk artifacts, experimental images of the apodized coded MLT schemes showed similar quality compared to that the non-coded schemes (Figures 8 and 11–13). It is important to note that the effective gain in frame rate of the Golay-coded schemes was compromised by a factor of two due to the transmission of complementary codes (Table 1).

Concerning spatial resolution, the coded schemes had slightly worse axial resolution whereas the lateral resolution was almost not affected (Figure 10); the latter, as is known, is affected by TT apodization. Moreover, as expected, the coded schemes had significantly improved the penetration, SNR, and contrast, especially at larger depths (Figures 11 and 12, Table 1). This was also expected given that more energy was transmitted into the tissue with the coded schemes given the increased time-bandwidth product [28]. It should be noticed that the transmit power of different imaging schemes was not normalized in order to make the different imaging schemes operating at their individual optimal conditions. Furthermore, the different blind regions of the experimental images are due to different signal lengths; indeed, longer excitations keep the transmitters switched on for longer times, during which the receivers are disabled.

Regarding crosstalk reduction, the tested Golay and chirp codes did not show much added value, and indeed, the apodization is still mandatory; the number of MLT beams is usually larger than the number of the orthogonal codes, hence the cross-talk cannot be perfectly suppressed; the frame rate might be decreased by a factor of two when complementary codes are needed. On the other hand, coded signals are beneficial in terms of SNR and penetration depth that are limited by MLT transmission schemes. In the latter, the pulse patterns transmitted to generate simultaneous beams, are given by the superimposition of different signals [7]. This superimposition leads to fully constructive interferences on the central elements. As such, the amplitudes applied on non-central elements would decrease since the maximal peak-peak voltage level is fixed on a given system, force a decrease of the amplitudes on the non-centered elements. As a consequence, less energy is transmitted, which would lower the penetration and SNR. The more the MLT beams, the lower the transmitted energy on a given MLT beam. This effect was well demonstrated in Figure 11a,b. In this perspective, coded excitation could thus be an alternative in maintaining the transmit energy/penetration in MLT imaging, particularly for schemes with a larger number of MLT beams. Moreover, a TT-apodized Golay-coded 4-MLT, 6-MLT, and a chirp-apodized 4-MLT could be good candidates as imaging scheme providing a good tradeoff between frame rate and signal-to-noise ratio. As a preliminary test, only one set of in-vivo images using Golay-coded 4-MLT and Chirp coded 4-MLT schemes were acquired. Further in-vivo investigation is the topic of the on-going work.

Finally, it should be noted that for either Golay or chirp codes, only two sets of orthogonal coded excitations were adopted since finding larger number of mutually orthogonal codes was challenging. Indeed, to have more mutually orthogonal Golay codes, the excitation duration would be significantly increased and/or more sets of complementary codes would be required [29,30]. Long excitation duration could lead to a large dead zone and strong interference of the MLT beams in the near field, whereas sending more complementary codes would result in no benefit in frame rate. As for chirp codes, more mutually orthogonal codes could be obtained by subdividing the transducer pass-band and by increasing the code duration. However, the transducer pass-band would not be effectively used, thus limiting the achievable range resolution. Moreover, a larger dead done could appear close to the transducer surface as well as more near-field artifacts due to higher interference between MLT beams with longer codes. Finally, stitching artifacts were seen in the in-vivo images as previously demonstrated in [9]. However, in the present study, this was not compensated, as described in Reference [31], since the primary purpose was to investigate the capability of coded excitation on crosstalk suppression.

5. Conclusions

The main goal of this study was to investigate the impact of orthogonal coded excitation for MLT crosstalk suppression. The results demonstrated that coded excitation could be used to suppress MLT crosstalk, although a Tukey-Tukey apodization is still required when the number of MLT beams is larger than the number of orthogonal codes. Moreover, coded excitation could help maintain the transmit energy (thus the SNR), which can be largely reduced as a result of increasing MLT beams. Overall, the benefit of applying coded excitation on an MLT imaging scheme to further suppress crosstalk artifacts remains limited, however when seeking a balance between frame rate and SNR, coded excitation could indeed be a valid alternative.

Supplementary Materials: The following are available http://www.mdpi.com/2076-3417/9/3/486/s1, Video S1: Figure13_a; Video S2: Figure13_b; Video S3: Figure13_c.

Author Contributions: Conceptualization, L.T., A.O., J.D.; Methodology, L.T., A.O.; Software, L.T.; Validation, L.T., Q.H., and A.R.; Writing-Original Draft Preparation, L.T.; Writing-Review & Editing, L.T., J.D., P.T., A.R., J.L., Q.H.; Visualization, L.T.; Supervision, J.D., J.L., P.T.; Funding Acquisition, L.T., A.R., J.L., J.D. and P.T.

Funding: The research leading to these results has received funding from China Postdoctoral Science Foundation (2014M560094), National Natural Science Foundation of China (81471665, 61871251, and 81561168023), the National Key R&D Program of China (2016YFC0102200 and 2016YFC0104700), European Research Council under the European Union's Seventh Framework Programme (FP7/2007-2013)/ERC Grant Agreement number 281748. A. Ramalli is supported by the European Union's Horizon 2020 research and innovation programme under the Marie Skłodowska-Curie grant agreement No 786027 (ACOUSTIC).

Conflicts of Interest: The authors declare no conflict of interest.

References

1. Cikes, M.; Tong, L.; Sutherland, G.R.; D'hooge, J. Ultrafast Cardiac Ultrasound Imaging. *JACC Cardiovasc. Imaging* **2014**, *7*, 812–823. [CrossRef] [PubMed]
2. Hasegawa, H.; Kanai, H. High-frame-rate echocardiography using diverging transmit beams and parallel receive beamforming. *J. Med. Ultrason.* **2011**, *38*, 129–140. [CrossRef]
3. Papadacci, C.; Pernot, M.; Couade, M.; Fink, M.; Tanter, M. High-contrast ultrafast imaging of the heart. *IEEE Trans. Ultrason. Ferroelectr. Freq. Control* **2014**, *61*, 288–301. [CrossRef] [PubMed]

4. Moore, C.; Castellucci, J.; Andersen, M.V.; Lefevre, M.; Arges, K.; Kisslo, J.; Von Ramm, O.T. Live high-frame-rate echocardiography. *IEEE Trans. Ultrason. Ferroelectr. Freq. Control* **2015**, *62*, 1779–1787. [CrossRef] [PubMed]
5. Montaldo, G.; Tanter, M.; Bercoff, J.; Benech, N.; Fink, M. Coherent plane-wave compounding for very high frame rate ultrasonography and transient elastography. *IEEE Trans. Ultrason. Ferroelectr. Freq. Control.* **2009**, *56*, 489–506. [CrossRef] [PubMed]
6. Shirasaka, T. Ultrasonic imaging apparatus. U.S. Patent US4815043, 1989.
7. Mallart, R.; France, L.C.; Fink, M. Improved imaging rate throught simultaneous transmission of several ultrasound beams. *SPIE* **1992**, *1733*, 120–130.
8. Tong, L.; Gao, H.; D'Hooge, J. Multi-transmit beam forming for fast cardiac imaging-a simulation study. *IEEE Trans. Ultrason. Ferroelectr. Freq. Control* **2013**, *60*, 1719–1731. [CrossRef]
9. Tong, L.; Ramalli, A.; Jasaityte, R.; Tortoli, P.; D'hooge, J. Multi-Transmit Beam Forming for Fast Cardiac Imaging—Experimental Validation and In Vivo Application. *IEEE Trans. Med. Imaging* **2014**, *33*, 1205–1219. [CrossRef]
10. Prieur, F.; Dénarié, B.; Austeng, A.; Torp, H. Multi-line transmission in medical imaging using the second-harmonic signal. *IEEE Trans. Ultrason. Ferroelectr. Freq. Control* **2013**, *60*, 2682–2692. [CrossRef]
11. Ramalli, A.; Dallai, A.; Guidi, F.; Bassi, L.; Boni, E.; Tong, L.; Fradella, G.; D'Hooge, J.; Tortoli, P. Real-Time High-Frame-Rate Cardiac B-Mode and Tissue Doppler Imaging Based on Multiline Transmission and Multiline Acquisition. *IEEE Trans. Ultrason. Ferroelectr. Freq. Control* **2018**, *65*, 2030–2041. [CrossRef]
12. Rabinovich, A.; Feuer, A.; Friedman, Z. Multi-line transmission combined with minimum variance beamforming in medical ultrasound imaging. *IEEE Trans. Ultrason. Ferroelectr. Freq. Control* **2015**, *62*, 814–827. [CrossRef] [PubMed]
13. Zurakhov, G.; Tong, L.; Ramalli, A.; Tortoli, P.; D'hooge, J.; Friedman, Z.; Adam, D. Multiline Transmit Beamforming Combined With Adaptive Apodization. *IEEE Trans. Ultrason. Ferroelectr. Freq. Control* **2018**, *65*, 535–545. [CrossRef] [PubMed]
14. Matrone, G.; Ramalli, A.; Savoia, A.S.; Tortoli, P.; Magenes, G. High Frame-Rate, High Resolution Ultrasound Imaging with Multi-Line Transmission and Filtered-Delay Multiply and Sum Beamforming. *IEEE Trans. Med. Imaging* **2016**, *PP*, 1–10. [CrossRef] [PubMed]
15. Chiao, R.Y.; Thomas, L.J. Synthetic transmit aperture imaging using orthogonal Golay coded excitation. In Proceedings of the 2000 IEEE Ultrasonics Symposium. Proceedings. An International Symposium, San Juan, Puerto Rico, 22–25 October 2000; pp. 1677–1680. [CrossRef]
16. Jeong, Y.K.; Song, T.-K. Simultaneous Multizone Focusing Method with Orthogonal Chirp Signals. In Proceedings of the IEEE Symposium on Ultrasonics, Atlanta, GA, USA, 7–10 October 2001; pp. 1517–1520.
17. Misaridis, T.; Jensen, J.A. Use of modulated excitation signals in medical ultrasound. Part III: High frame rate imaging. *IEEE Trans. Ultrason. Ferroelectr. Freq. Control* **2005**, *52*, 208–219. [CrossRef] [PubMed]
18. Chiao, R.Y.; Hao, X. Coded excitation for diagnostic ultrasound: A system developer's perspective. *IEEE Trans. Ultrason. Ferroelectr. Freq. Control* **2005**, *52*, 160–170. [CrossRef] [PubMed]
19. Kim, B.H.; Kim, G.D.; Song, T.K. A post-compression based ultrasound imaging technique for simultaneous transmit multi-zone focusing. *Ultrasonics* **2007**, *46*, 148–154. [CrossRef] [PubMed]
20. Yoon, C.; Yoo, Y.; Song, T.-K.; Chang, J.H. Orthogonal quadratic chirp signals for simultaneous multi-zone focusing in medical ultrasound imaging. *IEEE Trans. Ultrason. Ferroelectr. Freq. Control* **2012**, *59*, 1061–1069. [CrossRef]
21. Yoon, C.; Lee, W.; Chang, J.; Song, T.K.; Yoo, Y. An efficient pulse compression method of chirp-coded excitation in medical ultrasound imaging. *IEEE Trans. Ultrason. Ferroelectr. Freq. Control* **2013**, *60*, 2225–2229. [CrossRef]
22. Kim, B.-H.; Song, T.-K. Multiple transmit focusing using modified orthogonal Golay codes for small scale systems. In Proceedings of the IEEE Symposium on Ultrasonics, Honolulu, HI, USA, 5–8 October 2003; pp. 1574–1577. [CrossRef]
23. Jensen, J.A.; Svendsen, N.B. Calculation of Pressure Fields from Arbitrarily Shaped, Apodized, and Excited Ultrasound Transducers. *IEEE Trans. Ultrason. Ferroelectr. Freq. Control* **1992**, *39*, 262–267. [CrossRef]
24. OK-Jensen, J.A. FIELD: A Program for Simulating Ultrasound Systems. *Med. Biol. Eng. Comput.* **1996**, *34*, 351–352.

25. Tortoli, P.; Bassi, L.; Boni, E.; Dallai, A.; Guidi, F.; Ricci, S. ULA-OP: An Advanced Open Platform for Ultrasound Research. *IEEE Trans. Ultrason. Ferroelectr. Freq. Control* **2009**, *56*, 2207–2216. [CrossRef] [PubMed]
26. Boni, E.; Bassi, L.; Dallai, A.; Guidi, F.; Ramalli, A.; Ricci, S.; Housden, J.; Tortoli, P. A Reconfigurable and Programmable FPGA-Based System for Nonstandard Ultrasound Methods. *IEEE Trans. Ultrason. Ferroelectr. Freq. Control* **2011**, *59*, 1378–1385. [CrossRef] [PubMed]
27. Boni, E.; Cellai, A.; Ramalli, A.; Tortoli, P. A high performance board for acquisition of 64-channel ultrasound RF data. In Proceedings of the 2012 International Ultrasonics Symposium, Dresden, Germany, 5–7 October 2012; pp. 2067–2070. [CrossRef]
28. Ramalli, A.; Guidi, F.; Boni, E.; Tortoli, P. A real-time chirp-coded imaging system with tissue attenuation compensation. *Ultrasonics* **2015**, *60*, 65–75. [CrossRef] [PubMed]
29. Golay, M. Complementary series. *IEEE Trans. Inf. Theory* **1961**, *7*, 82–87. [CrossRef]
30. Tseng, C.C.; Liu, C.L. Complementary Sets of Sequences. *IEEE Trans. Inf. Theory* **1972**, *18*, 644–652. [CrossRef]
31. Denarie, B.; Bjastad, T.; Torp, H. Multi-line transmission in 3-D with reduced crosstalk artifacts: A proof of concept study. *IEEE Trans. Ultrason. Ferroelectr. Freq. Control* **2013**, *60*, 1708–1718. [CrossRef] [PubMed]

© 2019 by the authors. Licensee MDPI, Basel, Switzerland. This article is an open access article distributed under the terms and conditions of the Creative Commons Attribution (CC BY) license (http://creativecommons.org/licenses/by/4.0/).

Article

Spatial Coherence of Backscattered Signals in Multi-Line Transmit Ultrasound Imaging and Its Effect on Short-Lag Filtered-Delay Multiply and Sum Beamforming

Giulia Matrone [1,2,*] and Alessandro Ramalli [3,4]

1. Department of Electrical, Computer and Biomedical Engineering, University of Pavia, 27100 Pavia, Italy
2. Centre for Health Technologies, University of Pavia, 27100 Pavia, Italy
3. Laboratory on Cardiovascular Imaging and Dynamics, Department of Cardiovascular Sciences, KU Leuven, 3000 Leuven, Belgium; alessandro.ramalli@kuleuven.be
4. Department of Information Engineering, University of Florence, 50139 Florence, Italy
* Correspondence: giulia.matrone@unipv.it; Tel.: +39-0382-985918

Received: 6 February 2018; Accepted: 21 March 2018; Published: 23 March 2018

Abstract: In Multi-Line Transmission (MLT), high frame-rate ultrasound imaging is achieved by the simultaneous transmission of multiple focused beams along different directions, which unfortunately generates unwanted artifacts in the image due to inter-beam crosstalk. The Filtered-Delay Multiply and Sum (F-DMAS) beamformer, a non-linear spatial-coherence (SC)-based algorithm, was demonstrated to successfully reduce such artifacts, improving the spatial resolution at the same time. In this paper, we aim to provide further insights on the working principle and performance of F-DMAS beamforming in MLT imaging. First, we carry out an analytical study to analyze the behavior and trend of backscattered signals SC in MLT images, when the number of simultaneously transmitted beams and/or their angular spacing change. We then reconsider the F-DMAS algorithm proposing the "short-lag F-DMAS" formulation, in order to limit the maximum lag of signals used for the SC computation on which the beamformer is based. Therefore, we investigate in simulations how the performance of short-lag F-DMAS varies along with the maximum lag in the different MLT configurations considered. Finally, we establish a relation between the obtained results and the signals SC trend.

Keywords: filtered-delay multiply and sum beamforming; multi-line transmission; spatial coherence; ultrasound imaging

1. Introduction

Spatial coherence (SC) of ultrasound backscattered echoes has been the object of numerous works in the ultrasound imaging field. The first works date back to the 90s [1–5], when it was proposed to extend to pulse-echo ultrasound the Van Cittert Zernike (VCZ) theorem of statistical optics, which describes the spatial covariance of the wave field generated by an incoherent source [1].

Recently, the attention has been focused towards exploiting the concept of spatial coherence for the development of new image reconstruction and beamforming techniques. Some of the most renowned coherence-based methods are coherence-factor [6], phase-coherence and sign-coherence [7] beamforming, Filtered-Delay Multiply and Sum (F-DMAS) beamforming [8], and Short-Lag Spatial Coherence (SLSC) imaging [9]. The latter is a technique which directly generates an image of the spatial coherence of backscattered echoes evaluated at short lags, while all other methods aim at generating B-mode images, either via a coherence-based weighting or through non-linear correlation-like operations as in F-DMAS. Anyway, all these techniques use some estimate of spatial coherence to increase image quality. In general, they show enhanced clutter rejection and contrast improvement capabilities [9,10], which often also come along with an increase of lateral resolution [7,8]. For these reasons, in several studies, they have been proposed for use in application to techniques that partially sacrifice some of the above-mentioned image quality metrics to improve other crucial factors in ultrasound medical imaging, such as the frame rate. This is the case of well-known techniques like plane-wave imaging [11,12], multi-line acquisition (also called parallel beamforming) [13], and multi-line transmission (MLT) [14].

MLT, as the name suggests, is based on the use of multiple beams, which are transmitted simultaneously in the medium [15], reducing the acquisition time by a factor equal to the number of transmitted beams. It has recently gained a lot of interest, especially for application in cardiac ultrasound imaging [15], where the possibility to observe and track heart motion in real-time is of the uttermost importance for diagnostic purposes. In [15], it has been shown that, by applying a Tukey apodization window on both transmit and receive sides, improved performance can be achieved thanks to the lowering of cross-talk artifacts that arise when multiple beams are transmitted simultaneously, but at the expense of lateral resolution. Other works in the literature have addressed this same problem, such as [16–18].

Our group has recently worked towards the joint use of MLT with a new non-linear beamforming algorithm [19], i.e., F-DMAS [8,20,21], and showed how F-DMAS overcomes the main problems that have so far limited the use of MLT in clinical practice (i.e., cross-talk artifacts and low lateral resolution). This algorithm basically consists of computing the aperture spatial autocorrelation in reception; however, the pulse-echo response is heavily jeopardized by MLT imaging [22], hence the question we address in this work is: does this affect spatial coherence? If so, how?

To understand how spatial coherence varies in different imaging setups and with different image acquisition strategies can indeed be very useful, in that it allows us to explain the achievable performance and possible effects that could be observed in images generated with the previously mentioned spatial-coherence-based beamformers, which are obviously influenced by its behavior. In particular, here we are interested in analyzing the performance of F-DMAS when combined with different MLT strategies and when the spatial correlation operation, on which this algorithm is based, is only computed between signals within a certain maximum lag.

Therefore, the work presented here is organized in two parts. In the first part, an analytical study of the spatial coherence trend in a simulated uniform medium with MLT imaging is presented. Such analysis is performed by varying the main MLT parameters, i.e., the number of transmitted beams and/or their angular separation. In the second part, a modified version of F-DMAS beamforming, called short-lag F-DMAS, is introduced; the simulated results of MLT imaging with short-lag F-DMAS beamforming are analyzed in the different MLT scenarios mentioned above, also studying how the maximum spatial lag between signals that enter this beamformer influences image quality.

In the next sections, we will briefly recall the theoretical background of spatial coherence in ultrasound imaging and the main steps of the F-DMAS algorithm; the simulation setup will be described and the strategy used to study the trend of coherence with MLT will be presented (Section 2). Finally, the results of the coherence analytical study and F-DMAS performance in the considered test cases will be shown and discussed (Section 3). Conclusions will be provided in Section 4.

2. Materials and Methods

In Section 2.1, the spatial correlation concept is introduced and mathematically described, followed by a presentation of the MLT technique in Section 2.2. Subsequently, a description of short-lag F-DMAS beamforming is provided in Section 2.3. The simulation setup and the whole study organization are illustrated in Section 2.4.

2.1. Spatial Coherence

Spatial coherence (also called spatial correlation) is a measure of the coherence of the ultrasound beam reflected by a diffuse scattering medium. The VCZ theorem [2] states that spatial coherence can be obtained by computing the Fourier transform of the field intensity. Hence, in a classic B-mode scan, where dynamic focusing is only applied in reception, spatial coherence reaches its maximum value at the transmit focus, where the beam is narrower; moreover, the coherence function has a triangular shape, being the field a $sinc^2$ for a uniformly weighted aperture. Away from the focus, the beam widens and spatial coherence decreases. The same effect is observed in all those cases in which a decorrelation occurs, e.g., because of the presence of acoustic and/or electronics noise, beam sidelobes, and aberrations, etc. [9,20,23].

For each discrete time sample t, we define the normalized spatial covariance as:

$$C_N(l,t) = \frac{C(l,t)}{C(0,t)} = \frac{\sum_{n=1}^{N-l} \sum_{t=t_1}^{t_2} s_n(t) s_{n+l}(t)}{\sum_{n=1}^{N} \sum_{t=t_1}^{t_2} s_n^2(t)} \quad (1)$$

where $s_n(t)$ are the backscattered radio-frequency (RF) signals received by each n-th transducer of the N-element receive aperture and focused (delayed). l is an integer number representing the spatial lag, i.e., the number of elements between the couple of multiplied signals $s_n(t)$ and $s_{n+l}(t)$ ($l = 0 \ldots N - 1$), whose product is integrated over a short time interval $t = [t_1; t_2]$. $C(l,t)$ is the spatial covariance at the l-th lag; the normalization factor (denominator of Equation (1)) is given by the zero-lag covariance $C(0,t)$, as in [3].

Thus, a metric which accounts for the total spatial coherence $SC(t)$ (similar to what is proposed in [10] for SLSC imaging) can be computed as:

$$SC(t) = \sum_{l=0}^{N-1} C_N(l,t). \quad (2)$$

2.2. Multi-Line Transmission

MLT consists of transmitting N_{TX} beams simultaneously in different focusing directions. After each transmission (TX), the backscattered signals are collected by the array elements in reception (RX) and beamformed in parallel along the considered steering directions. Then, the beams are moved by an angular step $\theta_{STEP} = \theta_{SECT}/N_{LINES}$ in order to cover the full θ_{SECT}-wide image field of view with N_{LINES} scan lines. Therefore, the simultaneous TX beams are separated by an angular distance $\theta_{TX} = \theta_{SECT}/N_{TX}$ [22].

To achieve multiple beam TX, a set of focusing delays has to be computed for each of the simultaneous TX focusing directions. Thus, the MLT excitation pulse is obtained by summing up the excitation signals that would be used to focus the TX beam along each one of the considered steering directions in the classic single-line transmission (SLT) case.

As explained in [22], the main problem with MLT is the presence, in the final image, of the so called cross-talk artifacts due to interferences between the multiple beams.

To better understand how such artifacts generate, we have to recall the array theory [24], according to which the TX (h_{TX}) and RX (h_{RX}) ultrasound beams at the focal depth can be generally expressed as follows:

$$h_{TX}(u) = sinc(\frac{p_x u}{\lambda}) sinc[\frac{N p_x}{\lambda}(u - \sum_{j=-\infty}^{+\infty} \frac{j\lambda}{p_x} - u^{TX})] \quad (3)$$

$$h_{RX}(u) = sinc(\frac{p_x u}{\lambda}) sinc[\frac{N p_x}{\lambda}(u - \sum_{j=-\infty}^{+\infty} \frac{j\lambda}{p_x} - u^{RX})] \quad (4)$$

where $u = sin(\theta)$, u^{TX} refers to the focusing direction of the TX beam (θ^{TX}) and u^{RX} to the focusing direction of the RX beam (θ^{RX}). N is the number of array elements, p_x is the array pitch, and λ is the wavelength. Then, in the classic SLT case (i.e., each time one beam is transmitted and received), the pulse-echo (h_{PE}) response is given by:

$$h_{PE}(u) = h_{TX}(u) h_{RX}(u). \tag{5}$$

In MLT imaging, instead, several beams are transmitted in parallel, thus we have a summation in the TX beam expression:

$$h_{TX}(u) = sinc(\frac{p_x u}{\lambda}) \sum_{i=1}^{N_{TX}} sinc[\frac{N p_x}{\lambda}(u - \sum_{j=-\infty}^{+\infty} \frac{j\lambda}{p_x} - u_i^{TX})] \tag{6}$$

where u_i^{TX} refers to the focusing direction of the i-th TX beam (θ_i^{TX}). Hence, in this case the pulse-echo beam in (5) is made of N_{TX} terms, where one term is the one generated when $\theta_i^{TX} = \theta^{RX}$, i.e., the TX and RX responses are equal (as in a classic SLT scan), and the other terms with $\theta_i^{TX} \neq \theta^{RX}$ are the so called cross-talk contributions that originate from interferences among the TX and RX beams. Consequently, in the SLT case, when $N_{TX} = 1$, the pulse-echo beam has a $sinc^2$ shape, while in MLT with $N_{TX} > 1$, the beam has a more complex shape, with N_{TX} peaks in the simultaneous TX focusing directions, as shown in Figure 1 for MLT with four or 12 TX beams (i.e., 4-MLT and 12-MLT).

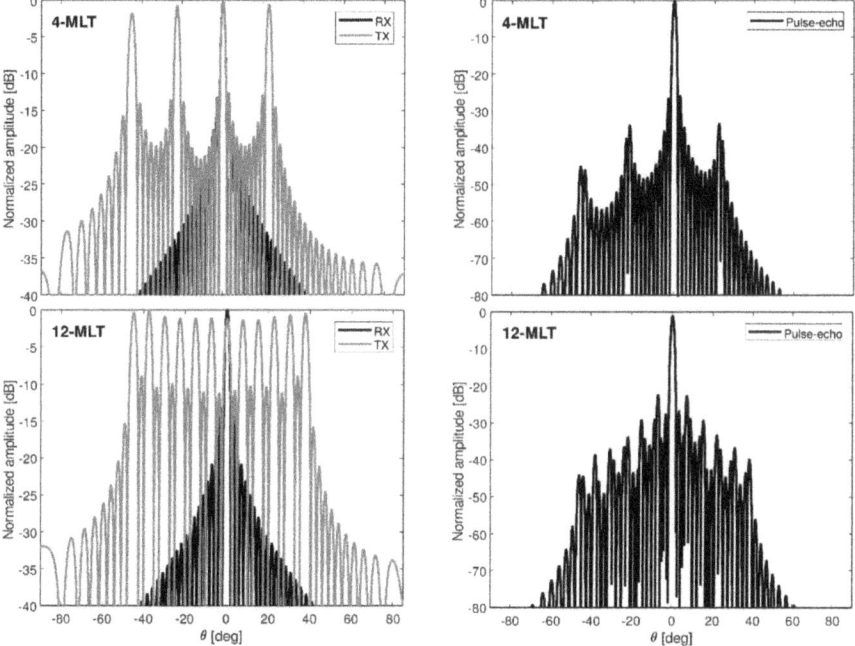

Figure 1. Examples of theoretical beampatterns in 4-MLT and 12-MLT: in the panels on the **left**, the TX (gray) and RX (black) beampatterns are plotted in each MLT case, when the RX beam is steered in one of the TX beams' direction (e.g., at 0°); on the **right**, the resulting pulse-echo beam shapes are shown.

In particular, interferences generate two types of cross-talk, i.e., the so called TX cross-talk, caused by interferences among TX beam side lobes and the RX beam main lobe, and RX cross-talk, caused by interferences between the TX beam main lobe and RX beam side lobes [22], which appear as artifacts around the beam main lobe along the axial or lateral direction, respectively.

2.3. Short-Lag Filtered-Delay Multiply and Sum Beamforming

F-DMAS beamforming derives from microwave imaging [25] and was recently introduced in ultrasound B-mode imaging by some of the authors [8]. This algorithm is based on the computation of the RX aperture spatial autocorrelation function. In F-DMAS, after focusing, the received RF signals $s_n(t)$ are amplitude-rescaled, by means of the signed square-root operator, and then combinatorially coupled and multiplied. The beamformed output is thus computed as:

$$y_{DMAS}(t) = \sum_{n=1}^{N-1} \sum_{m=n+1}^{N} sign(s_n(t)s_m(t)) \cdot \sqrt{|s_n(t)s_m(t)|}. \tag{7}$$

$y_{FDMAS}(t)$ is subsequently obtained by band-pass (BP) filtering $y_{DMAS}(t)$, in order to pass the second-harmonics that originates after signal cross-multiplications, and to attenuate as much as possible the baseband and higher frequency components which originate from such non-linear operations. After beamforming, the obtained RF image lines are demodulated using the Hilbert transform, normalized, and logarithmically compressed to produce the final B-mode image.

The F-DMAS formulation in (7) includes in the summation all possible couples among received RF signals, whose spatial lag $l = m - n$ varies from a minimum of 1 to a maximum M equal to $N-1$ (i.e., only auto-product terms are excluded). Since we are interested in studying the relation between SC and F-DMAS in MLT imaging, in this work we present the short-lag F-DMAS formulation, which limits the maximum lag M considered in the F-DMAS cross-multiplication stage. In particular, we will analyze the quality of images generated by using the following short-lag F-DMAS formulation:

$$y_{DMAS}(t, M) = \sum_{l=1}^{M} \sum_{n=1}^{N-l} sign(s_n(t)s_{n+l}(t)) \cdot \sqrt{|s_n(t)s_{n+l}(t)|}, \tag{8}$$

with $M = 1, 2, 3, \ldots, N-1$.

2.4. Simulation Setup and Study Organization

In this work, Matlab (The MathWorks, Natick, MA, USA) simulations with Field II [26,27] were carried out by modeling a 64-element phased array probe with a 340 µm pitch. The central working frequency was 2 MHz, and a 2-cycle Hanning-weighted sinusoidal burst was used as the excitation signal. The transmit focus was set at a 70 mm depth and dynamic focusing was implemented in reception. A 100 MHz sampling frequency was considered.

First, the analytical study on the spatial coherence trend was performed by simulating a numerical uniform speckle phantom, made of 140,000 scattering points with a Gaussian amplitude distribution, randomly placed in a $100 \times 1 \times 70$ mm^3 volume centered around $(x, y, z) = (0, 0, 65)$ mm (i.e., >10 scatterers per resolution cell).

In B-mode imaging with standard dynamic RX focusing, the spatial coherence is known to decrease before and after the fixed TX focus [20,23]; thus, in this work, both the normalized covariance $C_N(l,t)$ and total spatial correlation $SC(t)$ were evaluated using Equations (1) and (2), and averaging the values obtained in a small 5×5 mm^2 area centered at $(x, z) = (0, 70)$ mm, i.e., over the TX focus. Three different scenarios were analyzed (Table 1):

1. the number of MLT beams varies (i.e., $N_{TX} = 1/4/6/8/12$), but the total image sector is fixed ($\theta_{SECT} = 90°$), as well as the number of lines (192); consequently, the angle among the TX beams ($\theta_{TX} = \theta_{SECT}/N_{TX}$) changes together with the number of beams (usually, this is the classic MLT implementation);

2. the number of MLT beams varies (i.e., $N_{TX} = 1/4/6/8/12$), but the same angle (θ_{TX}) among the beams is used in all configurations; in particular, this angle was set to be equal to the one that would be obtained applying 12-MLT to scan a 90° sector (i.e., θ_{12}). Thus, in this case, the total image sector also changes in the different MLT configurations;
3. the number of beams is fixed (i.e., $N_{TX} = 4$), while the angle among them changes, as it would do in 4/6/8/12-MLT when a 90° sector is acquired. Thus, also here, the total image sector changes in each case.

Table 1. MLT configurations: number of TX beams and angular spacing.

Scenario 1		Scenario 2		Scenario 3	
N_{TX}	θ_{TX}	N_{TX}	θ_{TX}	N_{TX}	θ_{TX}
1	$\theta_1 = 90°$	1	θ_{12}	-	-
4	$\theta_4 = \theta_1/4$	4	θ_{12}	4	θ_4
6	$\theta_6 = \theta_1/6$	6	θ_{12}	4	θ_6
8	$\theta_8 = \theta_1/8$	8	θ_{12}	4	θ_8
12	$\theta_{12} = \theta_1/12$	12	θ_{12}	4	θ_{12}

These three scenarios were specifically chosen to evaluate the influence of the number of simultaneously transmitted beams or of their angular spacing, or of both factors, on the SC trend.

The second part of this study aimed at testing the performance of short-lag F-DMAS beamforming in MLT imaging. Results were also compared to those of DAS with Tukey apodization in RX. In both cases, Tukey apodization was also applied in TX, in order to reduce TX cross-talk [22].

Image quality was evaluated in terms of lateral resolution (at −6 dB), contrast ratio (CR), contrast-to-noise ratio (CNR), and speckle signal-to-noise ratio (sSNR) [8]:

$$CR = 20 log_{10}\left(\frac{\mu_{cyst}}{\mu_{bck}}\right), \qquad (9)$$

$$CNR = \frac{|\mu_{cyst} - \mu_{bck}|}{\sqrt{\sigma_{cyst}^2 + \sigma_{bckt}^2}}, \qquad (10)$$

$$sSNR = \frac{\mu_{bck}}{\sigma_{bck}}. \qquad (11)$$

Point spread function (PSF) simulations were carried out considering a series of vertically aligned point scatterers along the $x = 0$ mm axis, from $z = 10$ mm to 90 mm with a 20 mm step; in addition, resolution measurements were made at the focal depth on the PSF at $(x, z) = (0, 70)$ mm. CR, CNR, and sSNR were instead evaluated by simulating a numerical phantom of 140,000 points (i.e., >10 scatterers per resolution cell), with a 10-mm-diameter anechoic cyst centered at $(x, z) = (0, 70)$ mm, in a $100 \times 1 \times 70$ mm^3 uniform tissue background starting at $z = 30$ mm. Measurements were performed considering two 6-mm-diameter circular areas inside and outside the cyst.

3. Results and Discussion

3.1. Spatial Coherence Trends in MLT Images of a Homogeneous Phantom

Figure 2 shows the trend of backscattered signals SC (averaged over a small region around the focal depth, as said before), measured on the simulated uniform phantom images. The trend is plotted for SLT and 4/6/8/12 MLT in the three scenarios described in Section 2.4. Figure 3 represents the total spatial coherence values obtained by computing a summation over lags of the values plotted in Figure 2, for each MLT configuration (i.e., for each curve).

Figure 2 shows that SC has a triangular trend in SLT, which decreases from 1 at lag 0 to about -0.1 at lag 63. This trend is consistent with the one predicted by the VCZ theorem, which states that (at the focal depth) the SC of backscattered signals is the Fourier transform of the field intensity.

By looking at Figure 2a, we can also see that, when the number of TX beams increases (decreasing at the same time the angle among them), SC decreases more rapidly in the short-lag region, immediately after the peak at lag 0, and varies following a sort of damped oscillation for each different MLT case. On the whole, as shown by Figure 3a, total coherence decreases non-linearly as the number of MLT beams increases.

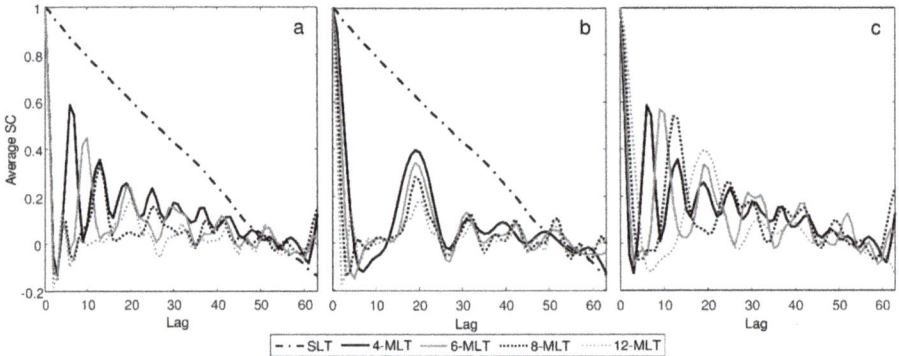

Figure 2. Normalized spatial covariance trend vs. lags, measured on the simulated uniform phantom data, and averaged over a 5×5 mm^2 area around $(x, z) = (0, 70)$ mm in the different scenarios of Table 1: (**a**) both the number of TX beams and the angle among them varies (as in SLT or 4/6/8/12-MLT); (**b**) only the number of TX beams varies (N_{TX} = 1/4/6/8/12), while the angular distance among them is fixed to the one of 12-MLT; (**c**) the number of TX beams is set to four, while the angular distance among them is the one of 4/6/8/12-MLT (in this last case, the legend refers to the number of TX beams which would determine the angular distance).

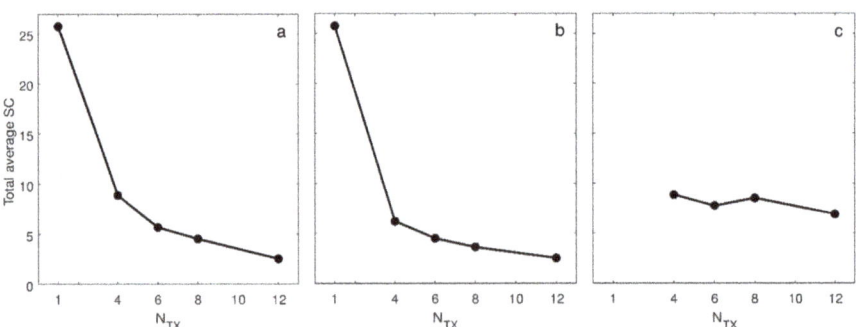

Figure 3. Total average spatial coherence, measured in a 5×5 mm^2 area around $(x, z) = (0, 70)$ mm in the different scenarios of Table 1: (**a**) both the number of TX beams and the angle among them varies (as in SLT or 4/6/8/12-MLT); (**b**) only the number of TX beams varies (N_{TX} = 1/4/6/8/12), while the angular distance among them is fixed to the one of 12-MLT; (**c**) the number of TX beams is set to four, while the angular distance among them is the one of 4/6/8/12-MLT (in this last case, the legend refers to the number of TX beams which would determine the angular distance).

In order to understand which factor determines such trends and an overall decrease of SC, we can look at Figure 2b,c and Figure 3b,c. Figure 2b refers to the case in which only the number of simultaneous TX beams varies, but the angle among them is fixed; in this way, it is possible to analyze the dependence of the SC trend on N_{TX} only. The plots clearly show that this time the SC oscillatory trend becomes similar for all MLT configurations (except for SLT), but the lobes have different peak amplitudes and widths, which still makes the total SC decrease when increasing the number of TX beams (Figure 3b). On the other hand, when the number of MLT beams N_{TX} is fixed, while the angle among them varies, we see in Figure 2c that the SC trend shows a different pitch between secondary lobes. Instead, total SC becomes similar in all cases (Figure 3c).

Therefore, by merging the information derived from the plots in Figures 2 and 3, we can observe that in all cases SC decreases when the lag between backscattered signals increases; this decreasing trend only follows a triangular pattern in the SLT case, when the pulse-echo beam is approximately a $sinc^2$, while a sort of damped oscillatory behavior is shown in MLT imaging, as expected because of the more complex beam shape obtained when multiple beams are transmitted simultaneously (cf. Figure 1). Furthermore, Figure 2b,c and Figure 3b,c demonstrate that it is the angular distance between the multiple TX beams (i.e., θ_{TX}) that mainly influences the SC trend over lags in MLT imaging, while the total SC value depends on the number of TX beams. When N_{TX} is fixed to four, for example, as in Figures 2c and 3c, total SC remains almost constant at ~8, while the lobes of the SC trend gradually move at higher lags with increasing values of N_{TX}. Conversely, when θ_{TX} is fixed (Figures 2b and 3b), we have a similar trend in all the MLT cases, with a peak at about lag 19, but the total SC decreases as N_{TX} becomes higher. Both behaviors can be observed in Figures 2a and 3a, where N_{TX} and θ_{TX} change at the same time in the different MLT configurations analyzed.

3.2. Simulated Images with MLT and Short-Lag F-DMAS

In this section, images generated with the short-lag F-DMAS formulation are analyzed, and the trend of each performance parameter is reported along with the maximum lag, in the MLT configurations considered.

First analyses have been performed on the simulated PSF at the focal depth, by measuring the lateral resolution at -6 dB; the results are plotted in Figure 4.

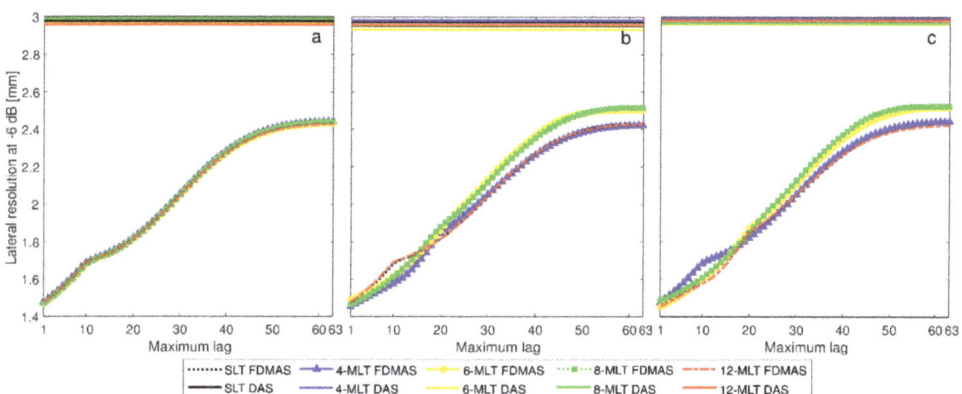

Figure 4. Lateral resolution (at -6 dB) trends, measured on the PSF images with varying maximum-lag, in the different scenarios of Table 1: (**a**) both the number of TX beams and the angle among them varies (as in SLT or 4/6/8/12-MLT); (**b**) only the number of TX beams varies ($N_{TX} = 1/4/6/8/12$), while the angular distance among them is fixed to the one of 12-MLT; (**c**) the number of TX beams is set to four, while the angular distance among them is the one of 4/6/8/12-MLT (in this last case, the legend refers to the number of TX beams which would determine the angular distance).

The plots in Figure 4a show that no significant difference among lateral resolution curves can generally be observed when the number of TX beams and their angular distance vary, from SLT to 4/6/8/12-MLT, both for DAS and F-DMAS images. In any case, a better resolution is always achieved by F-DMAS, with an increasing trend from ~1.4 mm at lag 1 to ~2.4 mm at lag 63 on average, as compared to ~3 mm on average for DAS.

When the angular distance between multiple TX beams is fixed (Figure 4b), the trend of F-DMAS image resolution remains almost the same, as observed in Figure 4a. The curves are actually slightly different in the 4/6/8/12-MLT cases, but no significant variation is observed: the trend is always increasing and better values are achieved by F-DMAS than DAS for all maximum-lag values.

Similar results are obtained in Figure 4c, when a fixed number (i.e., four) of TX beams is used while varying θ_{TX}.

Therefore, both DAS with Tukey RX apodization and F-DMAS are generally able to keep the lateral resolution unaltered when MLT with an increasing number of TX beams is applied to improve the frame-rate; this is probably because RX cross-talk in MLT mainly affects the lateral sides of the PSFs and not their main lobe width at −6 dB. However, F-DMAS both reduces RX cross-talk better than DAS and achieves up to a 1.5 mm better lateral resolution. If we consider the short-lag F-DMAS formulation, lateral resolution varies with the maximum lag employed; in particular, it worsens as the maximum lag increases.

As an example, Figure 5 (which refers to the 4-MLT configuration of scenario 1, cf. Table 1), clearly shows that the standard F-DMAS always achieves better RX cross-talk reduction than DAS, as well as lower side-lobe levels and a higher lateral resolution, which also makes TX cross-talk (that is suppressed in both cases by Tukey TX apodization) be confined to a narrower region [19]. When the maximum lag is reduced to 40 (Figure 5d) or 10 (Figure 5c), the PSF main-lobes become increasingly narrow as compared to the standard F-DMAS case (Figure 5b), but unfortunately, side-lobes and RX cross-talk artifacts become more visible at the same time (which, on the other hand, affects contrast performance, as will be shown hereinafter), becoming even worse than those of DAS when the maximum lag is limited to 10.

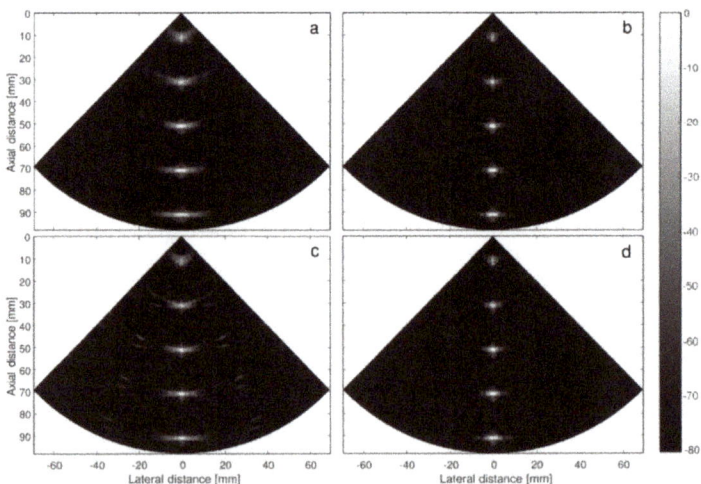

Figure 5. 4-MLT PSF images obtained in scenario 1 by applying (**a**) DAS with Tukey apodization in TX/RX, or F-DMAS with (**b**) maximum lag = 63 (standard version), (**c**) maximum lag = 10, (**d**) maximum lag = 40, and Tukey apodization in TX only. Figures are displayed over an 80 dB dynamic range, in order to better highlight the possible presence of small cross-talk artifacts.

This can be also observed by looking at Figure 6, which represents the lateral and axial profiles of the PSF at a 70 mm depth in the same four cases analyzed in Figure 5. In Figure 6a, we can see, for example, that RX cross talk artifacts (i.e., the two peaks at the left and right sides of the main lobe) are wider and also 15 dB higher using short-lag F-DMAS with maximum lag = 10, as compared to the standard F-DMAS case; on the other hand, the main lobe at −6 dB is narrower. The four axial sections of the PSFs in Figure 6b are instead very similar, also thanks to the application of Tukey TX apodization which lowers TX crosstalk in all cases [22].

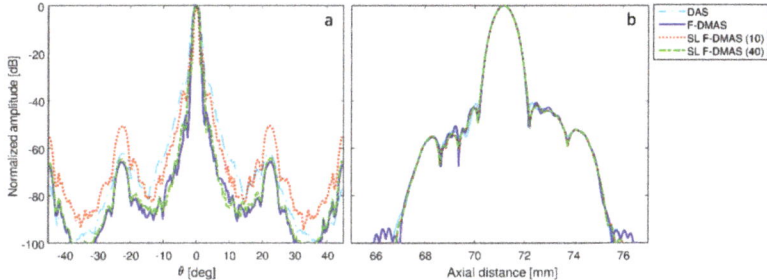

Figure 6. Lateral (**a**) and axial (**b**) profiles of the PSF at 70 mm (i.e., the TX focal depth) shown in Figure 5 for DAS, F-DMAS (standard formulation with maximum lag = 63), short-lag F-DMAS with maximum lag = 10 (SL F-DMAS (10)) and maximum lag = 40 (SL F-DMAS (40)).

The following results show how contrast and speckle quality are influenced by SC variations and by the maximum lag considered in short-lag F-DMAS.

Figure 7 represents the trend of DAS and F-DMAS image CR, CNR, and sSNR over maximum-lag values, measured on the cyst phantom in the three scenarios investigated.

By looking at the curves in Figure 7, we can see at first an increasing trend of the CR (in absolute value) together with the maximum lag employed in F-DMAS. When the classic F-DMAS formulation is used (maximum lag = 63), the CR is always better than that achieved in the corresponding DAS image, for all MLT cases (e.g., ~5–8 dB higher in scenario 1). This is not true in the short-lag region, where a maximum-lag threshold value exists, below which the CR of F-DMAS becomes worse than that of DAS. Such a threshold value is variable (generally in a range between lag 29 and 45) and seems to mainly depend on the angular distance between MLT beams (e.g., for 4-MLT in scenario 3 it is equal to 29 or 39 when θ_{TX} is the one that would be used in 4- or 12-MLT, respectively, cf. Figure 7CR-c). In scenario 2, this threshold is similar for all plots (i.e., lag = 39/41/42/42 for 4/6/8/12 TX beams, cf. Figure 7CR-b), since θ_{TX} is the same (i.e., θ_{12}). Moreover, panels CR-a and CR-c show that there is a dependence of the CR trend on the angular distance among TX beams, but no significant relation exists with the number of MLT beams (in panel CR-b in fact, all curves are almost overlapped). In particular, the CR gets worse as θ_{TX} decreases from θ_4 to θ_{12}.

For what concerns the CNR, the values obtained with F-DMAS are in any case lower than those achieved by DAS, as also previously reported [8,19]; their difference becomes higher when MLT is applied, i.e., the CNR at lag 63 with standard F-DMAS in scenario 1 is ~0.4 and ~0.8 (on average) lower than with DAS when implementing SLT and MLT, respectively.

The F-DMAS CNR exhibits a different trend compared to CR: it rapidly increases in the short-lag region (up to about lag 10), followed by a plateau up to lag 63, where the CNR is ~1.3 on average. Almost the same pattern can be observed in all the three scenarios and for all MLT configurations. By looking at Figure 7CNR-a, CNR-c (scenario 1 and 3), we can also see that the short-lag region trend depends on the angular distance between the multiple TX beams, where lower CNR values are obtained for smaller θ_{TX} values; on the other hand, when θ_{TX} is fixed (Figure 7CNR-b, scenario 2), the curves are almost overlapped. Thus, the CNR trend also seems to be related to θ_{TX}, even if in a less pronounced way as compared to the CR one.

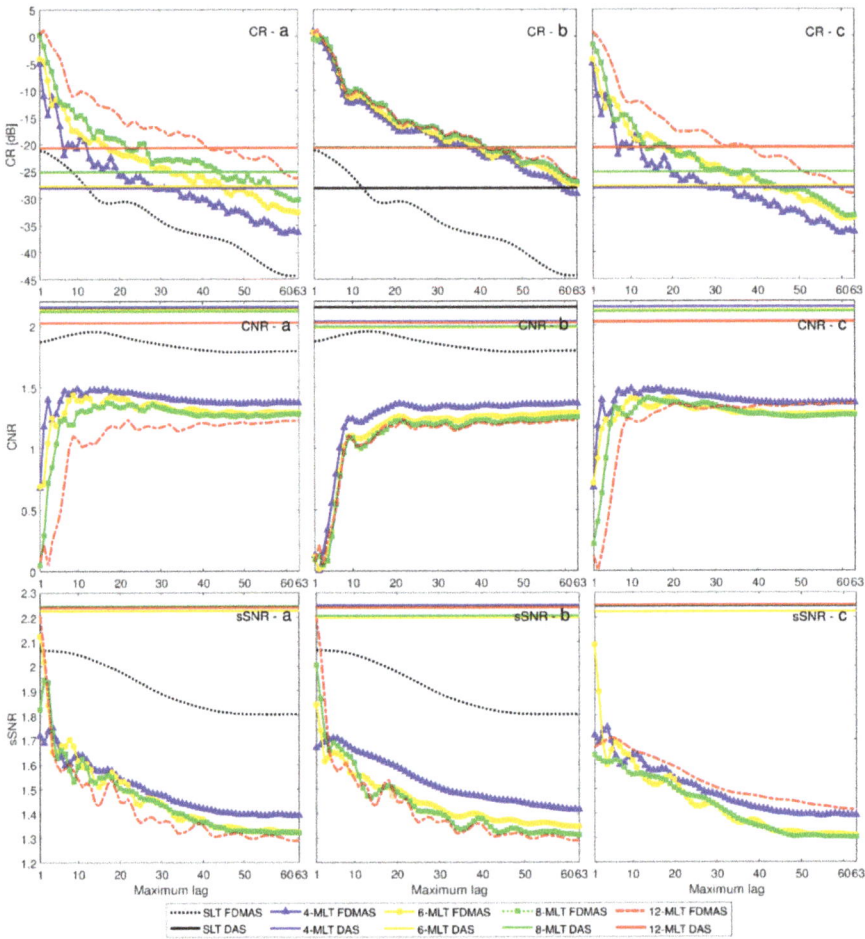

Figure 7. CR (**top** row), CNR (**middle** row), and sSNR (**bottom** row) trends, measured on the cyst-phantom images with varying maximum lag, in the different scenarios of Table 1: (**a**) both the number of TX beams and the angle among them varies (as in SLT or 4/6/8/12-MLT); (**b**) only the number of TX beams varies (N_{TX} = 1/4/6/8/12), while the angular distance among them is fixed to the one of 12-MLT; (**c**) the number of TX beams is set to four, while the angular distance among them is the one of 4/6/8/12-MLT (in this last case, the legend refers to the number of TX beams which would determine the angular distance).

Additionally, the F-DMAS sSNR is lower than that achieved by DAS, for all maximum-lag values and in all configurations; this is because F-DMAS impacts the speckle uniformity and a higher variance can be measured, as shown in [8]. As opposed to the CR and CNR trends, the sSNR decreases when the maximum lag increases; however, the total variation is quite small: it settles between 1.3 and 1.7 for lags ranging from 3 to 63, and the highest observable difference between the sSNR at lag 1 and at lag 63 is ~0.9 for 12-MLT in scenario 1 (Figure 7sSNR-a). Overall, for the sSNR trend, it is hard to clearly notice a dependence on the number of TX beams or their angular spacing, as in all scenarios, the curves are very similar.

Finally, to better highlight the short-lag F-DMAS contrast and sSNR performance, an example of images obtained in the classic 4-MLT configuration (i.e., scenario 1, Table 1) is presented in Figure 8. The figure shows an improved image quality achieved by F-DMAS as compared to DAS. Thanks to the higher lateral resolution, the cyst borders are always more clearly visible with F-DMAS, even when the maximum lag is limited to 10 and the CR is lower than that of DAS. The best resolution (~1.68 mm) is that achieved by the short-lag F-DMAS image with a maximum lag of 10 (Figure 8c); in this case, the cyst dimensions are almost equal to the real ones (i.e., 10 mm diameter), while for higher lags, it looks elliptic and smaller. When the maximum lag is 40 or 63 (Figure 8b,d), the lumen becomes increasingly dark as compared to the short-lag case in Figure 8c, and CR increases (about +9.3 dB and +15.5 dB in absolute value, respectively). This is further highlighted by Figure 9, which compares the cross-sections of the anechoic cysts in Figure 8, showing a definitely lower amplitude level for the pixels in the cyst lumen with standard F-DMAS, as compared to the case with maximum lag = 10.

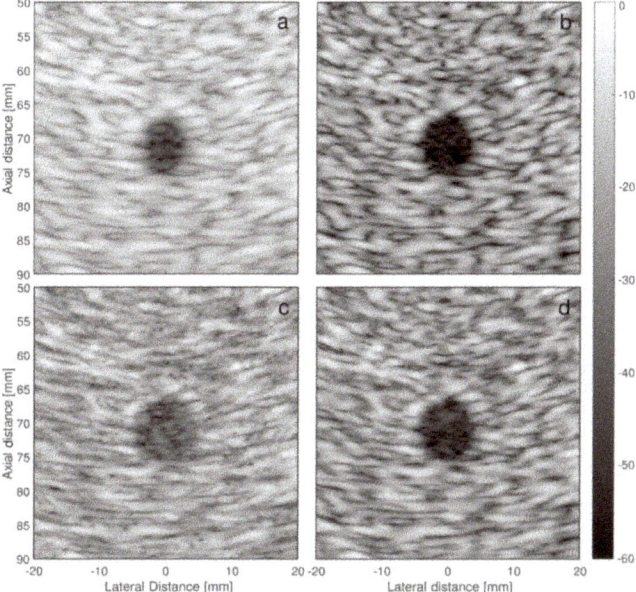

Figure 8. 4-MLT cyst-phantom images obtained in scenario 1 by applying (**a**) DAS with Tukey apodization in TX/RX, or F-DMAS with (**b**) maximum lag = 63 (full-version), (**c**) maximum lag = 10, (**d**) maximum lag = 40, and Tukey apodization in TX only. Figures are displayed over a 60 dB dynamic range.

The CNR is almost equal in Figure 8b,d (i.e., ~1.4), as well as the sSNR (i.e., ~1.4), and they are ~0.1 and ~0.25 lower than in Figure 8c, respectively.

On the whole, the results highlight a dependence of F-DMAS MLT image quality on backscattered signals SC from two points of view.

First, we have shown that, in MLT mode, performance varies with the number and angular spacing of simultaneously transmitted beams, which are in turn related to a variation of backscattered signals SC. The plots highlight that image contrast and speckle uniformity are mainly related to the angular distance between the multiple TX beams, and thus also to the trend of SC over lags; on the contrary, increasing the number of TX beams causes a drop of the measured total SC and has a lower impact on these parameters. In general, in fact, CR and CNR decrease when the angle between TX beams reduces, while the sSNR does not show an evident dependency on the number of TX beams or angular spacing. As expected, the lateral resolution measured at −6 dB is generally less or not influenced by the implemented MLT configuration.

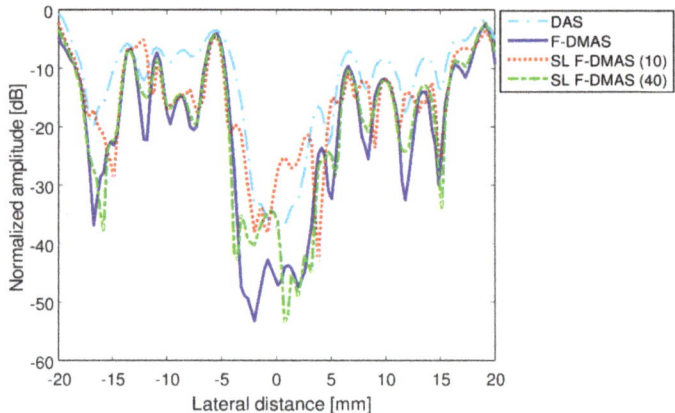

Figure 9. Cross-section of the anechoic cyst in Figure 8, for DAS, F-DMAS (standard formulation with maximum lag = 63), short-lag F-DMAS with maximum lag = 10 (SL F-DMAS (10)) and maximum lag = 40 (SL F-DMAS (40)).

Secondly, the results demonstrate that a relation exists between the analyzed outcome parameters and the maximum lag of received signals that enter the F-DMAS beamforming algorithm. Particularly, when only short-lag signal couples are used to generate the image, the PSF main lobe is narrower and the resolution is higher, but side-lobe and artifact suppression is not as good as in the full F-DMAS formulation case, which negatively affects image contrast. This is probably because in the short-lag region signals are highly correlated, being recorded by elements separated by a small distance in the probe, but the inclusion of higher-lag components in the F-DMAS summation provides further useful information for a better off-axis artifact rejection.

Albeit the presented analysis does not suggest an absolute criterion for the choice of an optimum maximum-lag value that could maximize all the analyzed performance parameters in F-DMAS images, it provides valuable indications to select a maximum-lag value (and thus a certain trade-off between contrast, resolution, and speckle SNR) based on the specific requirements of the considered application, also relating it to the frame-rate. For example, in those cases in which the contrast achieved by simple DAS could be sufficient but a higher resolution is required, like tissue Doppler applications [28], short-lag F-DMAS with low maximum-lag values can be employed; on the other hand, high maximum-lag values should be used when a high contrast is particularly required (e.g., for an effective segmentation of cardiac walls in ultrasound images [29]).

4. Conclusions

In this paper, we have presented an in-depth investigation of the performance of F-DMAS beamforming in the context of MLT imaging; this study is in fact not limited to the classic F-DMAS algorithm case, but it includes a thorough analysis of the image quality achievable with this beamformer when considering different maximum-lag values—and thus using the so called short-lag F-DMAS formulation—in different MLT imaging configurations.

The short-lag F-DMAS formulation provided new insights on the relation between the performance of the F-DMAS algorithm and the SC of backscattered signals. Simulation results show that the SC trend over lags and its total value are influenced by the number and angular spacing of simultaneous TX beams. This in turn influences the CNR, CR, sSNR, and resolution of F-DMAS images: higher maximum-lag values improve contrast, while lower values improve resolution and sSNR, which makes the quality of images change with the SC trend.

As a future development of this work, experimental tests will be carried out to assess the impact of noise on the trend of spatial correlation and on the image-quality parameters of interest considered here.

Acknowledgments: This work is supported by the University of Pavia, under the Blue Sky Research project MULTIWAVE. The authors would like to thank Alessandro Savoia (University Roma Tre) for all constructive discussions on the analysis of spatial coherence trends in MLT, and Piero Tortoli (University of Florence) and Giovanni Magenes (University of Pavia) for their insightful comments that helped improve the paper.

Author Contributions: G.M. conceived the study, designed the simulation tests, and worked on the processing of data; A.R. contributed to the design of the study and to simulation tests; G.M. wrote the paper; A.R. revised the manuscript. Both authors read and approved the submitted version of the paper.

Conflicts of Interest: The authors declare no conflict of interest.

References

1. Mallart, R.; Fink, M. The Van Cittert-Zernike theorem in pulsed ultrasound. Implications for ultrasound imaging. In Proceedings of the IEEE Ultrasonics Symposium, Honolulu, HI, USA, 4–7 December 1990; pp. 1603–1607.
2. Mallart, R.; Fink, M. Adaptive focusing in scattering media through sound-speed inhomogeneities: The Van Cittert Zernike approach and focusing criterion. *J. Acoust. Soc. Am.* **1994**, *96*, 3721–3732. [CrossRef]
3. Mallart, R.; Fink, M. The van Cittert–Zernike theorem in pulse echo measurements. *J. Acoust. Soc. Am.* **1991**, *90*, 2718–2727. [CrossRef]
4. Liu, D.-L.; Waag, R.C. About the application of the Van Cittert-Zernike theorem in ultrasonic imaging. *IEEE Trans. Ultrason. Ferroelectr. Freq. Control* **1995**, *42*, 590–601.
5. Li, P.-C.; Li, M.-L. Adaptive imaging using the generalized coherence factor. *IEEE Trans. Ultrason. Ferroelectr. Freq. Control* **2003**, *50*, 128–142. [PubMed]
6. Chen, J.F.; Zagzebski, J.A.; Dong, F.; Madsen, E.L. Estimating the spatial autocorrelation function for ultrasound scatterers in isotropic media. *Med. Phys.* **1998**, *25*, 648–655. [CrossRef] [PubMed]
7. Camacho, J.; Parrilla, M.; Fritsch, C. Phase coherence imaging. *IEEE Trans. Ultrason. Ferroelectr. Freq. Control* **2009**, *56*, 958–974. [CrossRef] [PubMed]
8. Matrone, G.; Savoia, A.S.; Caliano, G.; Magenes, G. The Delay Multiply and Sum beamforming algorithm in ultrasound B-mode medical imaging. *IEEE Trans. Med. Imaging* **2015**, *34*, 940–949. [CrossRef] [PubMed]
9. Lediju, M.A.; Trahey, G.E.; Byram, B.C.; Dahl, J.J. Short-Lag Spatial Coherence of backscattered echoes: imaging characteristics. *IEEE Trans. Ultrason. Ferroelectr. Freq. Control* **2011**, *58*, 1377–1388. [CrossRef] [PubMed]
10. Dahl, J.J.; Hyun, D.; Lediju, M.; Trahey, G.E. Lesion detectability in diagnostic ultrasound with short-lag spatial coherence imaging. *Ultrason. Imaging* **2011**, *33*, 119–133. [CrossRef] [PubMed]
11. Lu, J.-Y. 2D and 3D high frame-rate imaging with limited diffraction beams. *IEEE Trans. Ultrason. Ferroelectr. Freq. Control* **1997**, *44*, 839–856. [CrossRef]
12. Montaldo, G.; Tanter, M.; Bercoff, J.; Benech, N.; Fink, M. Coherent plane-wave compounding for very high frame rate ultrasonography and transient elastography. *IEEE Trans. Ultrason. Ferroelectr. Freq. Control* **2009**, *56*, 489–506. [CrossRef] [PubMed]
13. Shattuck, D.P.; Weinshenker, M.D.; Smith, S.W.; von Ramm, O.T. Explososcan: A parallel processing technique for high speed ultrasound imaging with linear phased arrays. *J. Acoust. Soc. Am.* **1984**, *75*, 1273–1282. [CrossRef] [PubMed]
14. Mallart, R.; Fink, M. Improved imaging rate through simultaneous transmission of several ultrasound beams. *Proc. SPIE* **1992**, *1773*, 120–130.
15. Tong, L.; Ramalli, A.; Jasaityte, R.; Tortoli, P.; D'hooge, J. Multi-transmit beam forming for fast cardiac imaging-Experimental Validation and in vivo application. *IEEE. Trans. Med. Imaging* **2014**, *33*, 1205–1219. [CrossRef] [PubMed]

16. Prieur, F.; Dénarié, B.; Austeng, A.; Torp, H. Multi-line transmission in medical imaging using the second-harmonic signal. *IEEE Trans. Ultrason. Ferroelectr. Freq. Control* **2013**, *60*, 2682–2692. [CrossRef] [PubMed]
17. Rabinovich, A.; Feuer, A.; Friedman, Z. Multi-line transmission combined with minimum variance beamforming in medical ultrasound imaging. *IEEE. Trans. Ultrason. Ferroelectr. Freq. Control* **2015**, *62*, 814–827. [CrossRef] [PubMed]
18. Zurakhov, G.; Tong, L.; Ramalli, A.; Tortoli, P.; D'hooge, J.; Friedman, Z.; Adam, D. Multi line transmit beamforming combined with adaptive apodization. *IEEE Trans. Ultrason. Ferroelectr. Freq. Control* **2018**. [CrossRef]
19. Matrone, G.; Ramalli, A.; Savoia, A.S.; Tortoli, P.; Magenes, G. High frame-rate, high resolution ultrasound imaging with multi-line transmission and Filtered-Delay Multiply and Sum beamforming. *IEEE Trans. Med. Imaging* **2017**, *36*, 478–486. [CrossRef] [PubMed]
20. Matrone, G.; Savoia, A.S.; Caliano, G.; Magenes, G. Depth-of-field enhancement in Filtered-Delay Multiply and Sum beamformed images using synthetic aperture focusing. *Ultrasonics* **2017**, *75*, 216–225. [CrossRef] [PubMed]
21. Matrone, G.; Ramalli, A.; Tortoli, P.; Magenes, G. Experimental evaluation of ultrasound higher-order harmonic imaging with Filtered-Delay Multiply And Sum (F-DMAS) non-linear beamforming. *Ultrasonics* **2018**, *86*, 59–68. [CrossRef] [PubMed]
22. Tong, L.; Gao, H.; D'hooge, J. Multi-transmit beam forming for fast cardiac imaging—A simulation study. *IEEE Trans. Ultrason. Ferroelectr. Freq. Control* **2013**, *60*, 1719–1731. [CrossRef] [PubMed]
23. Bottenus, N.; Byram, B.C.; Dahl, J.J.; Trahey, G.E. Synthetic Aperture Focusing for Short-Lag Spatial Coherence Imaging. *IEEE Trans. Ultrason. Ferroelectr. Freq. Control* **2013**, *60*, 1816–1826. [CrossRef] [PubMed]
24. Szabo, T.L. *Diagnostic Ultrasound Imaging: Inside Out*; Elsevier Academic Press: Hartford, CT, USA, 2004.
25. Lim, H.B.; Nhung, N.T.; Li, E.P.; Thang, N.D. Confocal Microwave Imaging for Breast Cancer Detection: Delay-Multiply-and-Sum Image Reconstruction Algorithm. *IEEE Trans. Biomed. Eng.* **2008**, *55*, 1697–1704. [PubMed]
26. Jensen, J.A.; Svendsen, N.B. Calculation of pressure fields from arbitrarily shaped, apodized, and excited ultrasound transducers. *IEEE Trans. Ultrason. Ferroelectr. Freq. Control* **1992**, *39*, 262–267. [CrossRef] [PubMed]
27. Jensen, J.A. Field: A program for simulating ultrasound systems. *Med. Biol. Eng. Comput.* **1996**, *34*, 351–353.
28. Tong, L.; Ramalli, A.; Tortoli, P.; Fradella, G.; Caciolli, S.; Luo, J.; D'hooge, J. Wide-Angle Tissue Doppler Imaging at High Frame Rate Using Multi-Line Transmit Beamforming: An Experimental Validation In Vivo. *IEEE Trans. Med. Imaging* **2016**, *35*, 521–528. [CrossRef] [PubMed]
29. Pedrosa, J.; Queiros, S.; Bernard, O.; Engvall, J.; Edvardsen, T.; Nagel, E.; D'hooge, J. Fast and Fully Automatic Left Ventricular Segmentation and Tracking in Echocardiography Using Shape-Based B-Spline Explicit Active Surfaces. *IEEE Trans. Med. Imaging* **2017**, *36*, 2287–2296. [CrossRef] [PubMed]

© 2018 by the authors. Licensee MDPI, Basel, Switzerland. This article is an open access article distributed under the terms and conditions of the Creative Commons Attribution (CC BY) license (http://creativecommons.org/licenses/by/4.0/).

Article

A Nonlinear Beamformer Based on *p*-th Root Compression—Application to Plane Wave Ultrasound Imaging

Maxime Polichetti [1,2,*], François Varray [1], Jean-Christophe Béra [2], Christian Cachard [1] and Barbara Nicolas [1]

[1] University Lyon, INSA-Lyon, UCBL, UJM-Saint-Etienne, CNRS, Inserm, CREATIS UMR 5220, U1206, F-69100 Villeurbanne, France; francois.varray@creatis.insa-lyon.fr (F.V.); christian.cachard@univ-lyon1.fr (C.C.); barbara.nicolas@creatis.insa-lyon.fr (B.N.)
[2] LabTAU, INSERM, Centre Léon Bérard, UCBL, University Lyon, F-69003 Lyon, France; jean-christophe.bera@inserm.fr
* Correspondence: maxime.polichetti@creatis.univ-lyon1.fr

Received: 23 February 2018; Accepted: 31 March 2018; Published: 11 April 2018

Abstract: Ultrafast medical ultrasound imaging is necessary for 3D and 4D ultrasound imaging, and it can also achieve high temporal resolution (thousands of frames per second) for monitoring of transient biological phenomena. However, reaching such frame rates involves reduction of image quality compared with that obtained with conventional ultrasound imaging, since the latter requires each image line to be reconstructed separately with a thin ultrasonic focused beam. There are many techniques to simultaneously acquire several image lines, although at the expense of resolution and contrast, due to interference from echoes from the whole medium. In this paper, a nonlinear beamformer is applied to plane wave imaging to improve resolution and contrast of ultrasound images. The method consists of the introduction of nonlinear operations in the conventional delay-and-sum (DAS) beamforming algorithm. To recover the value of each pixel, the raw radiofrequency signals are first dynamically focused and summed on the plane wave dimension. Then, their amplitudes are compressed using the signed p^{th} root. After summing on the element dimension, the signed *p*-power is applied to restore the original dimensionality in volts. Finally, a band-pass filter is used to remove artificial harmonics introduced by these nonlinear operations. The proposed method is referred to as *p*-DAS, and it has been tested here on numerical and experimental data from the open access platform of the Plane wave Imaging Challenge in Medical UltraSound (PICMUS). This study demonstrates that *p*-DAS achieves better resolution and artifact rejection than the conventional DAS (for $p = 2$ with eleven plane wave imaging on experimental phantoms, the lateral resolution is improved by 21%, and contrast ratio (CR) by 59%). However, like many coherence-based beamformers, it tends to distort the conventional speckle structure (contrast-to-noise-ratio (CNR) decreased by 45%). It is demonstrated that *p*-DAS, for $p = 2$, is very similar to the nonlinear filtered-delay-multiply-and-sum (FDMAS) beamforming, but also that its impact on image quality can be tuned changing the value of *p*.

Keywords: adaptive beamforming; p^{th} root; contrast enhancement; plane wave imaging; ultrasound imaging

1. Introduction

In ultrasound B-mode imaging, in terms of resolution and contrast, the image quality relies essentially on the strategy used to sonicate the biological tissue. The conventional focused approach consists of constructing the image lines one by one. Such approach suffers from particularly low frame rates (tens of frames per second) that are then not compatible with the present issues of ultrafast medical imaging allowing thousands of frames per second necessary for the the observation of transient biological phenomena [1], but also three-dimensional (3D), and even 4D, imaging [2].

To increase the frame rate, different sonication strategies have been investigated to decrease the number of acquisitions required to obtain a complete image. Multi-line acquisition uses wider beams for reconstruction, with several adjacent lines at the same time [3]. The multi-line transmit strategy consists of transmitting L simultaneous focused beams to reconstruct L image lines at the same time [4]. In both cases, the frame rate is multiplied by the number of simultaneous reconstructed lines. However, the image quality is impacted upon, due to interference from the echoes of different image lines.

To go further, all of the image lines can be reconstructed using the same set of radiofrequency (RF) signals after the whole medium has been sonicated with a single plane wave. In this case, the frame rate is multiplied by the number of image lines, which allows the acquisition of thousands of images per second. However, as less energy is sent into the medium and the echoes from the medium interfere, the resolution and the contrast are degraded. To limit this impact on image quality, several transmissions/receptions of steered plane waves can be combined to reconstruct a single image [5]. Montaldo et al. demonstrated that the image quality obtained with the conventional focused beam strategy can be recovered with plane wave compounding using a 10-fold greater frame rate. However, in the context of ultrafast imaging, the number of plane waves must be maintained as low as possible, and so image quality needs to be achieved with a complementary approach.

In addition, image quality relies on the beamformer used for reception, in order to process the raw echo signals into image lines. The conventional delay-and-sum (DAS) beamformer consists of correctly rephasing the raw signals acquired with the probe, and simply summing these. However, better resolution and contrast can be achieved by combining the delayed signals in a different way. The basic approach consists of applying a fixed weighting window before the sum of the delayed signals, to influence the shape of the point spread function (PSF). Typically, Tukey, Hann, or Hamming windows are used to reduce the side-lobe level, at the expense of a wider main lobe. Tong et al. investigated the influence of window shape on the rejection of interference during reception [4]. To go further, the adaptive Capon's minimum variance beamformer was proposed to compute the optimal weighting window that corresponds to each pixel [6,7]. This approach achieves a lot thinner main lobe and strong side-lobe rejection, compared with conventional DAS. Nevertheless, since finding the optimal set of weights for each pixel involves high computational complexity, the computing time required for beamforming is no more negligible than for simple DAS. A derived approach with lower complexity level was proposed by [8]. The optimal window for each pixel is chosen from a predefined set of windows. As the number of predefined windows increases, the image quality is enhanced, while the computational costs rise.

Other adaptive beamformers are instead based on coherence. These can reduce artifacts that originate from incoherent noise or interference. The coherence of N delayed samples is measured to obtain a weighting factor that is associated to each pixel value computed with the DAS beamformer. For instance, the generalized coherence factor is computed on the fast Fourier transform of the aperture [9], in order to define the coherent energy in the low frequencies. Alternatively, the phase coherence factor tends to reject the pixel value when the phase dispersion is high through the aperture [10]. The resolution and contrast enhancement, coupled with the low level of complexity (suitable for real-time implementation), make coherence approaches very attractive.

Recently, Matrone et al. proposed a nonlinear beamformer to enhance image quality, known as filtered-delay multiply-and-sum (FDMAS) [11], and demonstrated that this can be used to reject interference in the case of multi-line transmit imaging [12]. As this process correlates the N delayed samples, FDMAS can be considered as a beamformer that is based on coherence [13]. The value of each pixel is computed as the sum of the signed square roots of the corresponding N delayed samples, multiplied in pairs. The signed square root is used to keep the original pixel dimensionality in volts. However, these cross-products represent heavy computational costs.

A previous study demonstrated that the algebraic expression of FDMAS can be approximated as the squared sum of the signed square roots of the delayed samples [14]. Such formalism allows not only very similar performances of FDMAS to be recovered particularly rapidly, but also generalizes this approach to higher ranges by using the pth root. However, for even p-values, the sign of the oscillations is lost, which leads to the splitting of the frequency between the direct current (DC) components, and a doubling of the excitation frequency. This phenomenon tends to flatten the main lobe of the PSF compared to conventional DAS.

In the present study, a nonlinear beamformer is proposed, p-DAS, which consists of computing the value of each pixel as the signed p-power of the sum of the signed p^{th} root of the delayed samples. As the signs of the ultrasound oscillations are preserved through the algorithm, the frequency content is no longer split.

This paper is organized as follows. In the second section, the beamformers compared for this study are introduced in the context of ultrafast plane wave imaging, as DAS (conventional), FDMAS (nonlinear), and p-DAS (proposed method). Then, the numerical and experimental settings for the reconstruction of the Plane-Wave Imaging Challenge in Medical Ultrasound (PICMUS; IEEE IUS 2016) data [15] are described. In the third section, the principle and performance of p-DAS are illustrated. The last section concludes the paper and looks at several perspectives for the proposed method.

2. Methods and Materials

In this section, the three beamformers that are compared are presented in the context of plane wave compounding [5] (note that p-DAS beamforming could be applied to synthetic aperture or conventional focused imaging). Then, conventional DAS and FDMAS, as proposed by [11], and the here-proposed p-DAS method, are described. The data settings and reconstruction parameters are also given.

2.1. Methods

2.1.1. Conventional DAS for Plane Wave Imaging

In this section, the process for compounded plane wave imaging is described. The same linear ultrasound probe composed of N equally spaced elements is used for both transmission and reception. To reconstruct one image, M plane waves are transmitted. For a given plane wave m, with a transmission angle θ_m and a given element n of the array, the recorded signal is $p_{n,m}(t)$. The data acquired are beamformed to reconstruct the pixels in the grid (x, z), where x is the lateral axis parallel to the array, and z is the depth axis.

Each pixel $r(x, z)$ of the RF image is obtained through combination of the $N \times M$ correctly delayed samples $q_{n,m}(x, z)$. Each pixel is extracted from the raw echo signal $p_{n,m}(t)$. To correctly select the $q_{n,m}(x, z)$ samples, two assumptions are required. The speed of sound in tissues is believed to be constant $c_0 = 1540$ m·s^{-1}, and a point scatterer is believed to back-scatter the spherical wave front. In this way, the echo back-scattered by a point located at (x, z) corresponds to the samples $q_{n,m}(x, z)$ with:

$$q_{n,m}(x,z) = p_{n,m}(\tau_{n,m}(x,z)), \tag{1}$$

where $\tau_{n,m}(x, z)$ is the time of flight of the wave, which is the sum of $\tau_{TX,m}(x, z)$, the time in transmission for the steered plane wave m to get to the scatterer, and $\tau_{RX,n}(x, z)$, the time in reception from the scatterer to element n of the probe:

$$\tau_{n,m}(x,z) = \tau_{TX,m}(x,z) + \tau_{RX,n}(x,z), \tag{2}$$

$$\tau_{TX,m}(x,z) = \frac{1}{c_0}[x \cdot \sin \theta_m + z \cdot \cos \theta_m], \tag{3}$$

$$\tau_{RX,n}(x,z) = \frac{1}{c_0}\sqrt{(x-x_n)^2 + z^2},\quad (4)$$

where x_n is the lateral position of the nth element. Then, the $N \times M$ correctly delayed samples $q_{n,m}(x,z)$ are summed on the plane wave dimension, in order to obtain N compounded delayed samples $s_n(x,z)$:

$$s_n(x,z) = \sum_{m=1}^{M} q_{n,m}(x,z). \quad (5)$$

Finally, in the case of conventional DAS beamforming, the samples $s_n(x,z)$ are simply summed along the element dimension, in order to obtain the pixel value $r_{DAS}(x,z)$:

$$r_{DAS}(x,z) = \sum_{n=1}^{N} a_n \cdot s_n(x,z), \quad (6)$$

where a_n are the weighting coefficients of the apodization window (e.g., Tukey, Hann, and others). Finally, to obtain the B-mode image, the RF image is subjected to envelop detection along the depth dimension. Of note, if the central frequency of the ultrasound array is f_0, then the RF image lines oscillate around the spatial frequency $f_z = \frac{2}{c_0} f_0$. Thus, for the sake of simplicity, the parts that follow make direct use of the temporal frequency f_0 rather than the spatial frequency f_z.

In this study, all of the beamformers presented follow the same process for delaying and plane wave compounding. The only difference lies in the way $s_n(x,z)$ are combined to obtain the pixel value $r_{DAS}(x,z)$.

2.1.2. FDMAS Beamforming

For FDMAS beamforming, as proposed by [11], each pixel is reconstructed by multiplying the compounded delayed samples $s_n(x,z)$ in pairs. To keep the pixel dimensionality in volts, the signed square root is initially applied:

$$r_{FDMAS}(x,z) = \sum_{n=1}^{N-1} \sum_{n'=n+1}^{N} sign(s_n(x,z) \cdot s_{n'}(x,z)) \times \sqrt{|s_n(x,z) \cdot s_{n'}(x,z)|}. \quad (7)$$

The value of the pixel computed by FDMAS reflects an autocorrelation process for the receive aperture. In this way, FDMAS rejects noise and incoherent echoes with greater efficiency than DAS. Note that the multiplication of signals with the same polarity leads to the loss of the sign information. Therefore, if the RF signals $s_n(x,z)$ oscillate at f_0 (along the depth dimension), the spectrum of the image $r_{FDMAS}(x,z)$ is split between the DC component and $2f_0$. To retrieve narrow-band image lines before envelop detection, the RF image needs to be band-pass filtered at $2f_0$ along the depth dimension.

2.1.3. Proposed Method: p-DAS Beamforming

The proposed method is referred to as p-DAS. A block diagram is given in Figure 1 to describe how the value of a pixel $r_{p\text{-DAS}}(x,z)$ is computed from the corresponding compounded delayed samples $s_n(x,z)$ with the p-DAS beamformer. After being correctly delayed and compounded, the $s_n(x,z)$ amplitudes are compressed through the application of the signed p^{th} root:

$$\tilde{s}_n(x,z) = sign(s_n(x,z)) \cdot |s_n(x,z)|^{\frac{1}{p}}. \quad (8)$$

Note that the p-value can be an integer or a float. Then, the $\tilde{s}_n(x,z)$ are summed on the element dimension:

$$\tilde{r}_{p\text{-DAS}}(x,z) = \sum_{n=1}^{N} \tilde{s}_n(x,z). \quad (9)$$

With the original dimensionality of $s_n(x,z)$ in volts, this implies that $\tilde{r}_{p\text{-DAS}}(x,z)$ is homogeneous to volt$^{1/p}$. Thus, p-powering is necessary to recover the conventional image dimensionality in volts, as for DAS. Note that the signed p-power is used to keep the polarity of the samples. The value of the pixel obtained is then:

$$r_{p\text{-DAS}}(x,z) = \text{sign}(\tilde{r}_{p\text{-DAS}}(x,z)) \cdot |\tilde{r}_{p\text{-DAS}}(x,z)|^p. \tag{10}$$

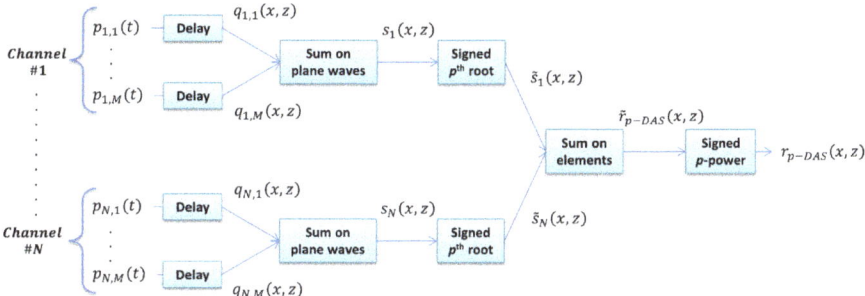

Figure 1. Block diagram describing how the value of a pixel $r_{p\text{-DAS}}(x,z)$ is obtained from the $N \times M$ raw echo signals recorded after M plane wave transmissions, using the N channels of the probe. Once all the pixel values have been computed in this way, a band-pass filter (not represented here) centered at f_0 is applied along the z-dimension of the complete image, in order to remove potential artificial harmonics due to nonlinear operations.

Note that while $s_n(x,z)$ oscillates at f_0 along the depth dimension, the nonlinear operations (such as signed p^{th} root and signed p-power) distort the shape of oscillations, and so generate artificial harmonics on $r_{p\text{-DAS}}(x,z)$, which have no acoustic meaning. The reconstructed image must be band-pass filtered to remove these artificial harmonics. The central frequency of this band-pass filter along the z-dimension is f_0. Note also that the frequency sampling has to be high enough (e.g., $\geq 8 f_0$) to ignore the potential aliasing of artificial harmonics.

As for FDMAS, p-DAS is a coherence-based nonlinear beamformer, but it has the advantage that it is tunable with the p-value to balance the effects of the beamformer on the images. Moreover, p-DAS preserves the sign of the oscillations through the reconstruction process thanks to the use of the signed p^{th} root and the signed p-power. Such difference and the impact on the images are investigated in Section 3.3.

In actuality, p-DAS can be viewed as DAS applied to the signed p^{th} root of the raw signals, followed by the signed p-power, in order to re-establish the original dimension of the image. In this way, the effects of the coherent summation are reinforced. The principle of the proposed method is illustrated in Section 3.1, considering p-DAS as a DAS beamformer with adaptive weighting, as in [6]. Indeed, Equation (9) can be rearranged as:

$$\tilde{r}_{p\text{-DAS}}(x,z) = \sum_{n=1}^{N} w_n(x,z) \cdot s_n(x,z), \tag{11}$$

where $w_n(x,z)$ are the adaptive weights that depend on the $s_n(x,z)$ amplitudes:

$$w_n(x,z) = \frac{1}{|s_n(x,z)|^{\frac{p-1}{p}}}. \tag{12}$$

2.2. Materials

The performances of DAS, FDMAS, and *p*-DAS are compared based on the data acquired for the four phantoms available on the open access PICMUS platform (IEEE IUS 2016) [15]. The PICMUS challenge was specifically developed for a challenge on advanced beamforming methods using quantitative criteria for image quality. In this way, resolution and contrast are evaluated separately on two numerical phantoms and two experimental phantoms, using Matlab (R2016b, The MathWorks, Natick, MA, USA). The probe settings and the parameters for transmission are given in Table 1. The performances are evaluated with single plane wave imaging (1 PW), and also with 11 plane wave imaging (11 PW), uniformly tilted from $\theta_1 = -10°$ to $\theta_M = +10°$.

Table 1. Transmission settings for the PICMUS challenge [15].

Number of Elements N	128	Excitation (number of cycles)	2.5
Pitch (mm)	0.3	Transmit frequency $f_0 = f_s/4$ (MHz)	5.2
Element width (mm)	0.27	Sampling Frequency f_s (MHz)	20.8

2.2.1. Beamforming Parameters

This part describes the details and parameters used for the three beamformers compared: DAS, FDMAS, and *p*-DAS. For all of these beamformers, the raw RF data are previously oversampled at $2f_s = 41.6$ MHz (corresponding to $8f_0$). For this paper, the over sampling frequency was empirically determined to ignore the potential aliasing of artificial harmonics when using *p*-DAS. A dynamic aperture with a constant F-number of 1.75 is used, as by [5]. For DAS, uniform receive apodization is used. For FDMAS, the band-pass filter used is the same as that described by [16], as a Kaiser finite impulse response (FIR) filter with a centred frequency at $2f_0$ and a pass-band defined on the range $1.5f_0 - 2.5f_0$. For *p*-DAS, two *p*-values are investigated: $p = 2$ and $p = 3$. The band-pass filter required to remove artificial harmonics on the image is performed along the depth dimension. The RF image is first low-pass filtered using a Butterworth filter (order 11; cut-off frequency $1.7f_0$), and then it is high-pass filtered using another Butterworth filter (order 11; cut-off frequency $0.4f_0$).

2.2.2. Image Quality Metrics

To evaluate image quality, the metrics are computed for the envelop images (before log-compression). The mean resolutions in the axial and lateral directions are automatically measured as described in [15]: the full width at half maximum (FWHM) is averaged over the 20 point scatterers for the numerical phantom (Figure 2a–d), and over seven point scatterers embedded in the speckle for experimental phantom (Figure 2i–l). This averaging is necessary as the FWHM is not spatially constant [17]. The contrast is measured according to two criteria: the mean contrast ratio (CR), and the mean contrast-to-noise ratio (CNR):

$$CR = \frac{1}{K}\sum_{k=1}^{K} 20.\log_{10}\left(\frac{\mu_{speckle,k}}{\mu_{cyst,k}}\right), \qquad (13)$$

$$CNR = \frac{1}{K}\sum_{k=1}^{K} 20.\log_{10}\left(\frac{|\mu_{speckle,k} - \mu_{cyst,k}|}{\sqrt{\frac{\sigma^2_{speckle,k} + \sigma^2_{cyst,k}}{2}}}\right), \qquad (14)$$

where *k* is the index of the cyst. $\mu_{speckle,k}$ (respectively $\mu_{cyst,k}$) is the mean pixel amplitude in the speckle ring (respectively inside the kth cyst). $\sigma^2_{speckle,k}$ (respectively $\sigma^2_{cyst,k}$) is the variance of the pixel amplitude in the speckle ring (respectively inside the kth cyst). For simulation, the phantom is composed of $K = 9$

cysts (Figure 2e–h), and, for the experiment, the phantom is composed of $K = 2$ cysts (Figure 2m–p). The full details of this dataset are available in [15].

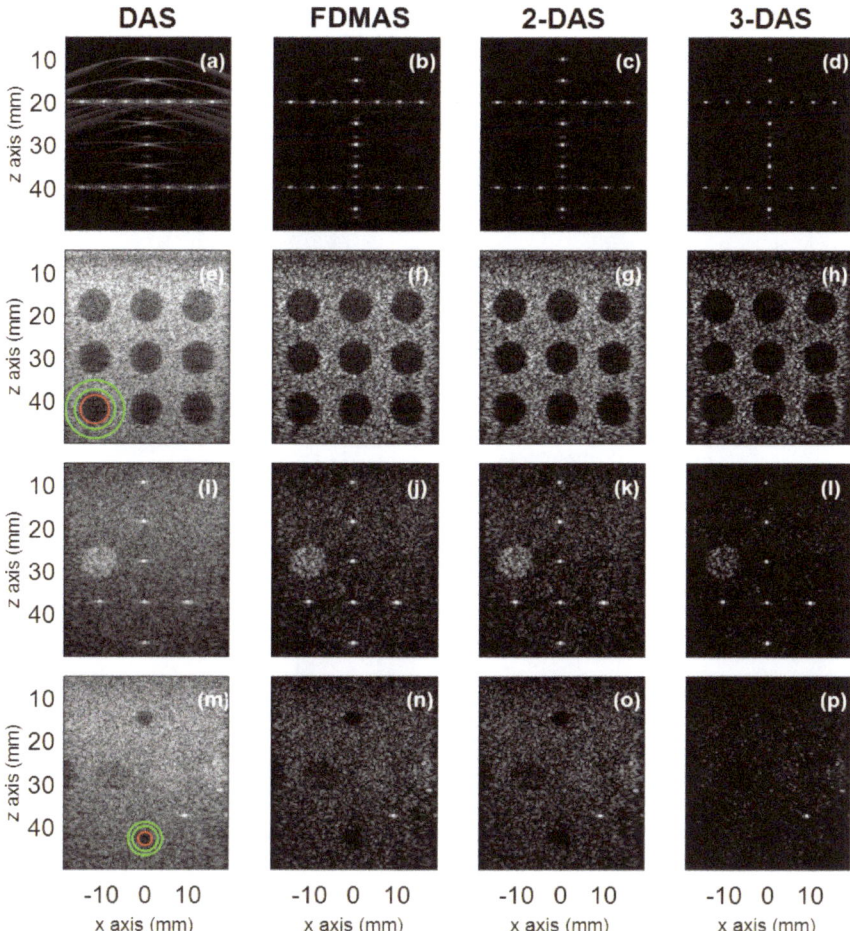

Figure 2. B-mode images obtained with single plane wave imaging for the four phantoms and the four beamformers compared: DAS (**a,e,i,m**), FDMAS (**b,f,j,n**), 2-DAS (**c,g,k,o**), and 3-DAS (**d,h,l,p**). (**a–d**) the numerical phantom used for the resolution; (**e–h**) the numerical phantom used for the contrast; (**i–l**) the experimental phantom used for the resolution; and (**m–p**) the experimental phantom used for the contrast. All of the images are displayed with a 60-dB dynamic range. An example of the boundaries chosen for the contrast metrics is given for simulation (**e**) and experiment (**m**), with the inside of the cyst in red, and the outside of the speckle ring in green.

3. Results and Discussion

In this section, the three beamformers (i.e., DAS, FDMAS, p-DAS) are analyzed and compared through the results obtained in simulation and experiments. In the first subsection, p-DAS is illustrated with a simulated point target. Then, the performances of the p-DAS in terms of the image quality (resolution and contrast) for the four phantoms of the PICMUS challenge are presented, for 1 PW and 11 PW. The last subsection makes the comparison of images reconstructed with FDMAS and p-DAS beamformers.

3.1. Analysis on the Principle of p-DAS

First of all, the principle of the p-DAS beamformer is compared to the DAS beamformer, for the case of single plane wave imaging. Namely, their respective PSFs (shown in Figure 3a,b) are investigated in simulation on the resolution phantom described by [15]. Here, p-DAS is considered as a DAS beamformer with adaptive weighting, considering Equations (11) and (12).

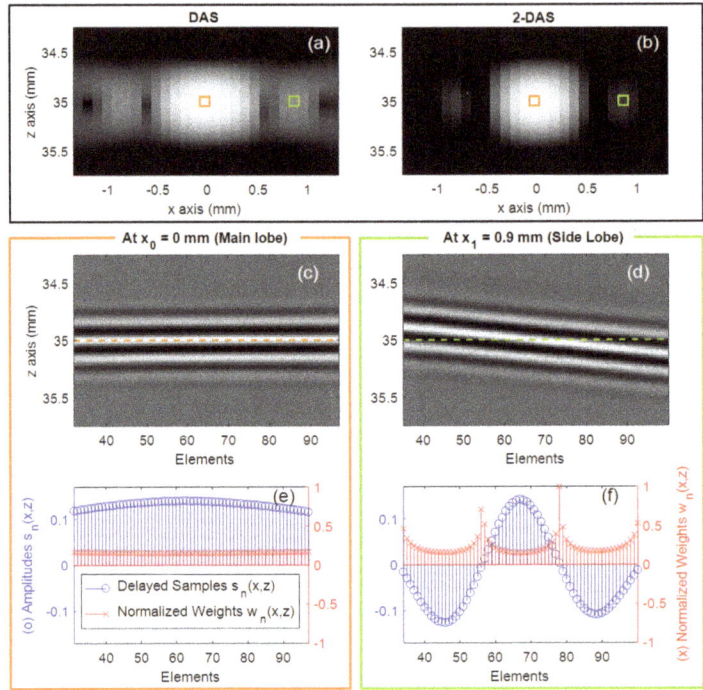

Figure 3. Illustration of the enhanced side-lobe rejection for the point spread function (PSF) with the proposed method with $p = 2$ (2-DAS), compared with conventional DAS. The scatterer placed at ($x_0 = 0$ mm, $z_0 = 35$ mm) on the numerical phantom is considered. The B-mode log-compressed images with a 40-dB dynamic range are shown for DAS (**a**) and 2-DAS (**b**). The delayed signals $s_n(x,z)$ are shown for $x_0 = 0$ mm (**c**) and $x_1 = 0.9$ mm (**d**); (**e**) the delayed samples $s_n(x_0, z_0)$ for the pixel on the main lobe are shown in blue, with their corresponding adaptive weights $w_n(x_0, z_0)$ in red; and (**f**) the delayed samples $s_n(x_1, z_0)$ for the pixel on the side lobe are shown in blue, with their corresponding adaptive weights $w_n(x_1, z_0)$ in red.

For the point scatterer placed at ($x_0 = 0$ mm, $z_0 = 35$ mm), the PSF obtained with 2-DAS is shown in Figure 3b, and this shows better side-lobe rejection compared with the PSF obtained with DAS, as shown in Figure 3a. To explain this observation, the two beamformers are compared through consideration of two specific pixels: the maximum of the main lobe at ($x_0 = 0$ mm, $z_0 = 35$ mm), which corresponds to the point scatterer location, and the peak side-lobe at ($x_1 = 0.9$ mm, $z_0 = 35$ mm). First, the pixel reconstruction at the main lobe is investigated. In Figure 3c, the oscillations are successfully rephased for the pixel at (x_0, z_0) because a wave front actually comes from this location. As a result, a uniform level of amplitude is observed for $s_n(x_0, z_0)$ in Figure 3e (blue). For DAS, these delayed samples are simply summed. For 2-DAS, an adaptive weighting window is applied (Figure 3e, red). However, the N weighting values $w_n(x_0, z_0)$ are quasi-identical, as they depend on the $s_n(x_0, z_0)$ amplitudes, which are similar along the elements. Finally, this pixel reconstructed with 2-DAS is equivalent to that reconstructed with DAS.

Then, DAS and 2-DAS are compared for the pixel at ($x_1 = 0.9$ mm, $z_0 = 35$ mm), which corresponds to the peak side-lobe position for the reconstruction with DAS. The value of the pixel should tend to 0, as no scatterer is present. However, it is corrupted with the energy of the ill-rephased wave front in Figure 3d. As a result, the corresponding $s_n(x_1, z_0)$ amplitudes are no longer uniform (Figure 3f, blue). To lower the value of the pixel reached with conventional DAS, the adaptive weights $w_n(x_1, z_0)$ of 2-DAS tend to be the strongest for the lowest amplitudes, as demonstrated in Equation (12). This is why side-lobes vanish for 2-DAS (Figure 3b) compared to DAS (Figure 3a).

Note that, considering Equation (12), the side-lobe rejection increases with the p-value, as stronger weights are applied. The following subsection presents the impact on the B-mode images that arises from this adaptive approach.

3.2. Performances Evaluation of p-DAS on Image Quality

In this subsection, the image qualities obtained with the DAS and p-DAS ($p = 2$ and $p = 3$) beamformers are investigated, in order to understand the influence of the value of p on images. The images obtained for single plane wave imaging (1 PW), and 11 plane wave imaging (11 PW) are presented and analyzed.

3.2.1. Single Plane Wave Imaging

First, the results obtained for single plane wave imaging (1 PW) are considered. The images obtained on the four phantoms with 1 PW are shown in Figure 2. The measurements are summed in Figure 4. As a first observation, the resolution is enhanced with p-DAS, due to its better rejection of ill-rephased wave fronts: in Figure 4, as p increases, improved sensitivity in the lateral direction is highlighted, both in simulation (Figure 4b; DAS, 0.73 mm; 2-DAS, 0.53 mm; 3-DAS, 0.46 mm) and experiment (Figure 4f; DAS, 0.81 mm; 2-DAS, 0.58 mm; 3-DAS, 0.48 mm).

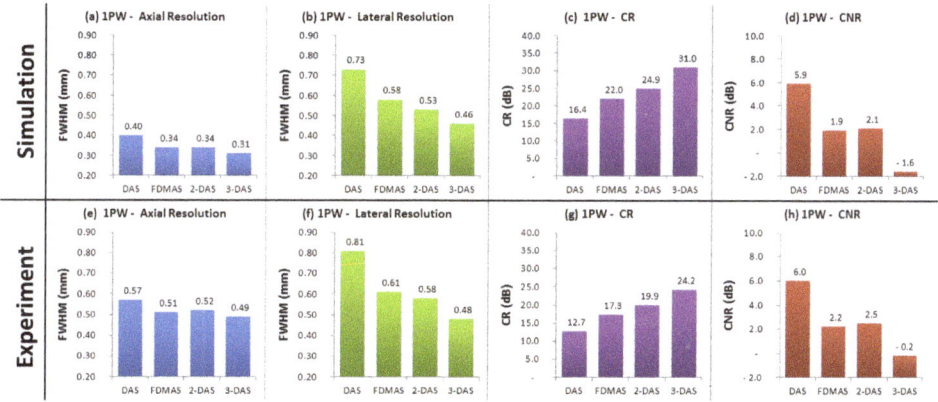

Figure 4. Performances obtained with DAS, FDMAS, 2-DAS, and 3-DAS beamformers, using a single plane wave in simulations (**a–d**) and experiments (**e–h**). The results obtained are given for axial resolution (**a,e**), lateral resolution (**b,f**), contrast ratio (**c,g**), and contrast-to-noise ratio (**d,h**).

The mean axial FWHM over the 20 targets relies mainly on the excitation waveform, and it is slightly decreased for $p = 2$ and $p = 3$ (Figure 4a,e). For the simulated phantom (Figure 2a,c), the 20 axial FWHM are quasi constant for all of the scatterers with DAS (0.39–0.41 mm), but they decrease for the interfering scatterers when using 2-DAS (0.31–0.41 mm). The target that is most affected by interference located at ($x = 0$ mm, $z = 20$ mm) has an axial FWHM of 0.40 mm with

DAS, whereas this is 0.31 mm with 2-DAS. However, the FWHM of the isolated scatterer placed at ($x = 0$ mm, $z = 10$ mm) is 0.40 mm for DAS and 2-DAS. Indeed, the rejection of interference for p-DAS is nonlinear with respect to amplitude, as seen for Equation (12). Then, the same interference noise is not rejected in the same way whether it interferes with the top of the pulse, or with the low-amplitude edges of the pulse. Finally, p-DAS tends to shrink the axial FWHM in the presence of interference.

In addition, the results demonstrate that the CR increases with the p-value for both simulation (Figure 4c; DAS, 16.4 dB; 2-DAS, 24.9 dB; 3-DAS, 31.0 dB) and experiment (Figure 4g; DAS, 12.7 dB; 2-DAS, 19.9 dB; 3-DAS, 24.2 dB). These observations confirm the conclusions of Section 3.1: the higher the p-value is, the more rejected the side-lobes are. As a result, the artifacts from the bright speckle into the dark cysts are attenuated for $p = 2$ (Figure 2g,o), and even more so for $p = 3$ (Figure 2h,p), with respect to those for DAS (Figure 2e,m).

Conversely, the CNR decreases as the p-value increases, both for simulation (Figure 4d; DAS, 5.9 dB; 2-DAS, 2.1 dB; 3-DAS, -1.6 dB) and experiment (Figure 4h; DAS, 6.0 dB; 2-DAS, 2.5 dB; 3-DAS, -0.2 dB). This means that the variance of the pixel intensities inside the cysts and inside the speckle is stronger with $p = 2$ and even more with $p = 3$. Looking at the simulated images, the speckle obtained with DAS (Figure 2e,m) is relatively homogeneous with gray pixels, whereas for $p = 2$ (Figure 2g,o) and $p = 3$ (Figure 2h,p) the speckle is more heterogeneous, with a background that is darker. Indeed, the coherent bright spots in the speckle are not strongly impacted by p-DAS, whereas the incoherent dark pixels of the speckle are heavily rejected. As a result, the CNR drops for high p-values because it varies in the opposite way to the increased variance. The same trend is obtained for the experiment (Figure 2m,o,p).

Moreover, roughly coherent wave fronts back-scattered from point targets appear as relatively brighter spots when using $p = 2$ or $p = 3$ (Figure 2k,l), rather than conventional DAS (Figure 2i). Note that the energy of such high coherent targets tends to darken the speckle in their near lateral neighborhood. This phenomenon was identified by Ole et al. in [18] as an inner characteristic of beamformers based on coherence. For this reason, the p-value can be adjusted to enhance the resolution and the CR (useful for lesion detectability), while preserving the speckle structure and CNR (used for texture analysis, and so, tissue characterization) [19]. Such a trade-off was identified as common behavior for adaptive beamformers by [20].

3.2.2. Eleven Plane Wave Imaging

The results for 11 PW imaging are analyzed in this section. The images obtained on the four phantoms with 11 PW are shown in Figure 5. The measurements are summarized in Figure 6. The results demonstrate that p-DAS can be applied successfully to plane wave compounding since the effects of p-DAS obtained with 1 PW are preserved when using 11 PW (i.e., as p increases, better lateral resolution and CR, but worse CNR). The lateral resolution is improved as p increases both in simulation (Figure 6b; DAS, 0.62 mm; 2-DAS, 0.49 mm; 3-DAS, 0.44 mm) and experiment (Figure 6f; DAS, 0.65 mm; 2-DAS, 0.51 mm; 3-DAS, 0.44 mm). The CR is also improved for higher p-values, for both simulation (Figure 6c; DAS, 28.1 dB; 2-DAS, 43.7 dB; 3-DAS, 55.1 dB) and experiment (Figure 6g; DAS, 21.0 dB; 2-DAS, 33.6 dB; 3-DAS, 41.4 dB). In the image (Figure 5e,g,h), the cysts are darker with respect to the speckle when using $p = 2$ or $p = 3$. This can be explained as follows. If the compounding leads to different mean values for each channel, this is because no coherent source is detected, but only noise or interference: the nonlinear weighting attenuates the noisy pixel value. In this way, the interferences from bright speckles in the cysts vanish, so the cysts tend to be darker, and so the CR is improved. As p-DAS increases the gap between coherent and incoherent sources, the variance inside the cyst and inside the speckle are also increased. Such darkening of the speckle structure is also noticeable on experimental phantoms (Figure 5m,o,p). In this way, the CNR decreases with the p-values for both simulation (Figure 6d; DAS, 6.5 dB; 2-DAS, 3.0 dB; 3-DAS, 0.1 dB) and experiment (Figure 6g; DAS, 7.7 dB; 2-DAS, 4.1 dB; 3-DAS, 1.3 dB). Finally, the impact of p-DAS on the axial FWHM is less than with 1 PW, for both simulation (Figure 6a; DAS, 0.40 mm; 2-DAS, 0.39 mm; 3-DAS, 0.38 mm) and experiment (Figure 6e; DAS, 0.56 mm; 2-DAS, 0.55 mm; 3-DAS, 0.54 mm). Indeed, as observed for

1 PW, the more a scatterer is impacted by interference, the more its FWHM is improved. However, as the compounding of the 11 PW before p-DAS attenuates the level of interference, the nonlinear behavior of p-DAS with respect to the amplitude is less highlighted.

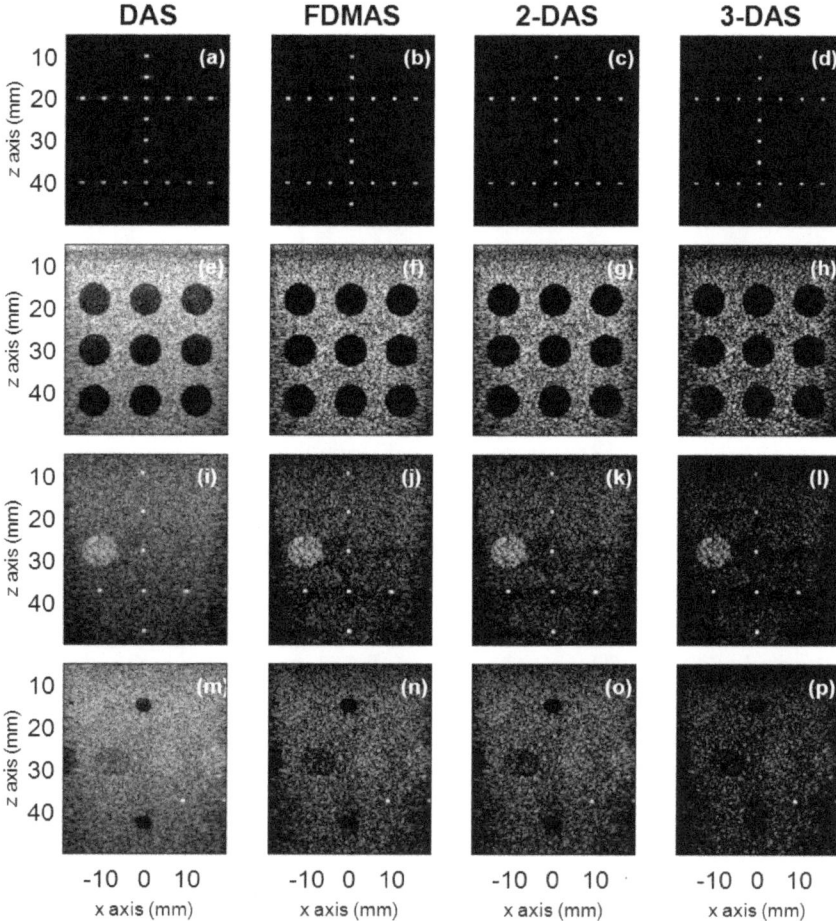

Figure 5. B-mode images obtained with eleven plane wave imaging for the four phantoms and the four beamformers compared: DAS (**a,e,i,m**), FDMAS (**b,f,j,n**), 2-DAS (**c,g,k,o**), and 3-DAS (**d,h,l,p**). (**a–d**) the numerical phantom used for the resolution; (**e–h**) the numerical phantom used for the contrast; (**i–l**) the experimental phantom used for the resolution; and (**m–p**) the experimental phantom used for the contrast. All of the images are displayed with 60-dB dynamic range.

In summary, p-DAS is an interesting beamformer to reject any incoherent pixel values that result from either ill rephased wavefronts, or waveforms corrupted by interference or noise. This explains the better lateral resolution and enhanced side-lobe rejection, and so the improved contrast ratio. These three first metrics increase with the p-value. However, the CNR is degraded, as the speckle results from interference, and so it is heavily rejected with respect to the coherent target. Finally, p-DAS appears to be a good candidate to resolve particularly punctual targets embedded in speckle or noise.

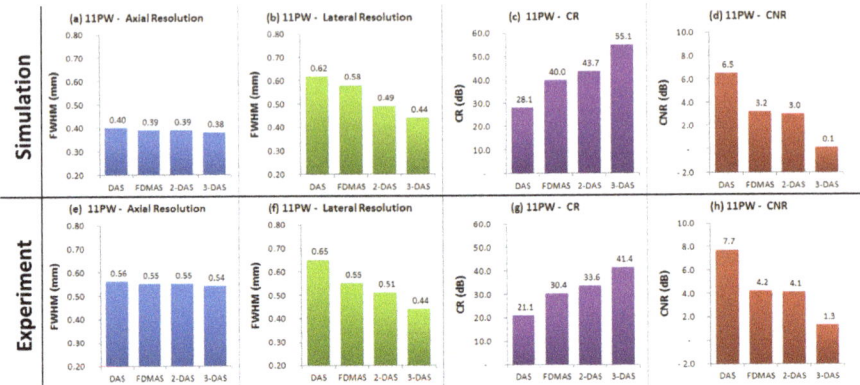

Figure 6. Performances obtained with DAS, FDMAS, 2-DAS, and 3-DAS beamformers, using 11 plane waves in simulation (**a**–**d**) and experiment (**e**–**h**). Results obtained are given for axial resolution (**a**,**e**), lateral resolution (**b**,**f**), contrast ratio (**c**,**g**), and contrast-to-noise ratio (**d**,**h**).

3.3. Comparison between FDMAS and p-DAS

The comparison between 2-DAS and FDMAS is seen through the results obtained with single plane wave imaging. First, the images obtained with a single plane wave on the numerical phantom are compared (see Figure 2b,c). Namely, the PSFs obtained with 2-DAS and FDMAS (Figure 7a,b) are compared for the point scatterer placed at ($x_0 = 0$ mm, $z_0 = 35$ mm). Their normalized lateral sections are plotted in Figure 7c. As a first observation, the main lobe is not flat any more when using 2-DAS beamforming rather than FDMAS beamforming. This explains the better lateral resolution for 2-DAS with respect to FDMAS (Figure 4b; 2-DAS, 0.53 mm; FDMAS, 0.58 mm). This enhancement is also seen in the experiment (Figure 4f; 2-DAS, 0.58 mm; FDMAS, 0.61 mm). Moreover, the relative enhanced side-lobe rejection of 2 dB observed for PSFs in Figure 7c leads to slightly better CR for 2-DAS than for FDMAS, for both simulation (Figure 4c; 2-DAS, 24.9 dB; FDMAS, 22.0 dB) and experiment (Figure 4g; 2-DAS, 19.9 dB; FDMAS, 17.3 dB). To explain these better performances, the role of the 'signed' p-power in p-DAS rather than the 'unsigned' p-power is highlighted. A previous study by [14] demonstrated that images obtained with 2-DAS with unsigned p-power or with FDMAS are equivalent. Indeed, considering Equation (10) with the unsigned p-power with $p = 2$ leads to:

$$\tilde{r}_{unsigned,\text{2-DAS}}(x,z) = \left[\sum_{n=1}^{N} \tilde{s}_n(x,z)\right]^2. \tag{15}$$

Then, the rearranged algebraic expression of Equation (15) gives:

$$\underbrace{\tilde{r}_{unsigned,\text{2-DAS}}(x,z)}_{\text{ⓐ } p\text{-DAS with unsigned } p\text{-power}} = \underbrace{\sum_{n=1}^{N} \tilde{s}_n^2(x,z)}_{\text{ⓑ Sum of } |s_n(x,z)|} + 2 \times \underbrace{\sum_{n=1}^{N-1} \sum_{n'=n+1}^{N} \tilde{s}_n(x,z)\tilde{s}_{n'}(x,z)}_{\text{ⓒ FDMAS}}. \tag{16}$$

Finally, FDMAS expression is recovered in Equation (16), where the term (b) is negligible compared to the term (c). Indeed, when these originate from the same wave front, the $s_n(x,z)$ have almost the same amplitudes, whatever the index n. This means that, in Equation (16), $\tilde{s}_n(x,z)^2$ and $\tilde{s}_n(x,z)\tilde{s}_{n'}(x,z)$ are particularly close values. For Equation (16), as there are only N terms in the sum (b) but $\frac{N(N-1)}{2}$ in the sum (c), the sum (b) can be neglected, and then $\tilde{r}_{unsigned,\text{2-DAS}}(x,z)$ is approximatively equal to $r_{FDMAS}(x,z)$. Note that a band-pass filter centred at $2f_0$ for the RF images is necessary when using 2-DAS with the unsigned p-power, as the sign information is lost and thus the frequency content is split between the DC component and $2f_0$. However, using p-DAS (i.e., with the signed p-power) preserves the sign of the oscillations, and thus avoids splitting the frequency content of RF images. In this way, the band-pass filter selects the entire information maintained at f_0, instead of just keeping the reduced part split at $2f_0$, as for FDMAS.

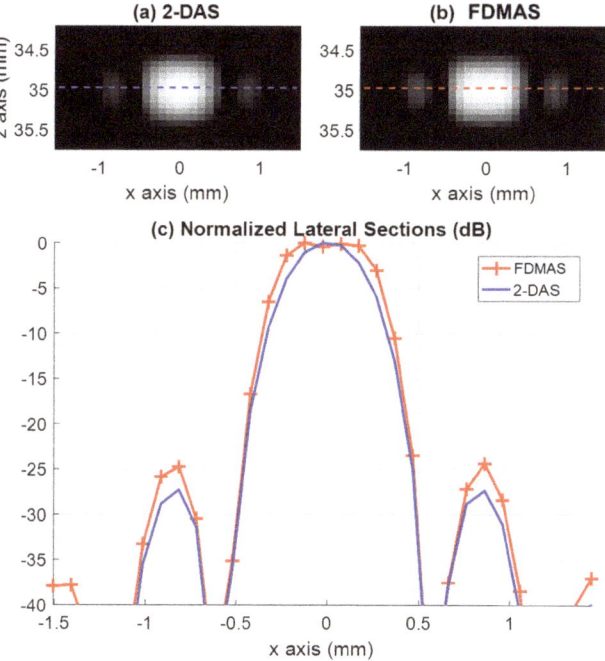

Figure 7. Comparison of PSFs obtained with 2-DAS and FDMAS. The scatterer was placed at ($x_0 = 0$ mm, $z_0 = 35$ mm) on the numerical phantom. B-mode log-compressed images are shown over a 40-dB dynamic range for 2-DAS (**a**) and FDMAS (**b**). Their respective normalized lateral profiles are shown in (**c**).

In summary, compared to FDMAS, 2-DAS avoids the flatenning of the main lobe and allows a slight better side-lobe rejection of 2 dB (Figure 7), which is consistent with the slight improved lateral resolutions and contrast ratios on the different numerical and experimental phantoms. Moreover, with the generalized formalism of p-DAS, the proposed beamformer can be tuned with the p-value to adjust the trade-off between the CR and the CNR. Finally, 2-DAS has a similar formulation as the real-time implementation of FDMAS proposed by Ramalli et al. [21].

4. Conclusions

This paper proposes a nonlinear beamformer, p-DAS, to enhance image quality. This beamformer is tested here in simulation and also under experimental conditions in the context of ultrafast plane wave imaging. The main benefits of this method are to improve lateral resolution, to better reject side lobes, and, as a consequence, to improve the contrast ratio, depending on the p-value used. However, p-DAS tends to distort the speckle statistics as the nonlinear operations increase the gap between the coherent and incoherent targets that compose the speckle, which leads to decreased CNR.

Another interesting aspect is that the proposed method is an extended version of an already well-known nonlinear beamformer: FDMAS. The signed p-power allows the removal of the PSF flat main lobe in FDMAS and better side-lobe rejection. Finally, the p-value can be tuned to find the best trade-off between CR and CNR, as required by the user.

Several applications are expected for this p-DAS beamforming. First, as p-DAS highlights punctual targets in noisy environments, it appears to be a very promising tool for bubble localization. For instance, Errico et al. reconstructed high-resolution vascular maps using ultrafast imaging of microbubbles [22]. The efficiency of the method relied on the assumption that the bubbles are punctual and separable sources, and then on the detection of a large number of them in B-mode images. p-DAS also represents a solution to increase the number of detected bubbles, as when the echo signals are weak due to ultrasound attenuation (e.g., deep tissues, through the skull, high frequency ultrasound). Another application might be imaging of cavitation for therapeutic monitoring. In [23], Boulos et al. proposed enhancing the resolution of passive imaging of cavitation with the adaptive phase coherence factor based beamformer [10]. p-DAS might also be an appropriate tool to reject interference between the punctual bubbles that compose the cavitation cloud, which generate strong artifacts on cavitation maps. Moreover, the performance of p-DAS might be investigated with other high frame rate imaging strategies. In the case of multi-line transmit [4,12], p-DAS might be a good candidate to reject the cross-talk for reception.

For the methodology, further studies can be conducted on the function used to compress the adaptive weights. Indeed, the coupling of 'p^{th}-root and p-power' has been used in this paper, while it might be interesting to use other couplings of functions, such as 'exponential and logarithm', to understand their impact on image quality.

Acknowledgments: The authors would like to thank Giulia Matrone and Alessandro Stuart Savoia for their valuable discussions, and for sharing their experience of FDMAS. This work was supported by LABEX CELYA (ANR-10-LABX-0060) and was performed within the framework of the LABEX PRIMES (ANR-11-LABX-0063) of the Université de Lyon, within the programme '*Investissements d'Avenir*' (ANR-11-IDEX-0007), operated by the French National Research Agency (ANR).

Author Contributions: Maxime Polichetti, François Varray and Barbara Nicolas developed the methodology, analyzed the data, and wrote the paper. Christian Cachard and Jean-Christophe Béra are the leaders of the project, and contributed with reading and improving the manuscript.

Conflicts of Interest: The authors declare that they have no conflicts of interest.

References

1. Tanter, M.; Fink, M. Ultrafast imaging in biomedical ultrasound. *IEEE Trans. Ultrason. Ferroelectr. Freq. Control* **2014**, *61*, 102–119.
2. Petrusca, L.; Varray, F.; Souchon, R.; Bernard, A.; Chapelon, J.Y.; Liebgott, H.; N'Djin, W.A.; Viallon, M. Fast Volumetric Ultrasound B-Mode and Doppler Imaging with a New High-Channels Density Platform for Advanced 4D Cardiac Imaging/Therapy. *Appl. Sci.* **2018**, *8*, 200.
3. Tong, L.; Gao, H.; Choi, H.F.; D'hooge, J. Comparison of conventional parallel beamforming with plane wave and diverging wave imaging for cardiac applications: A simulation study. *IEEE Trans. Ultrason. Ferroelectr. Freq. Control* **2012**, *59*, 1654–1663.
4. Tong, L.; Gao, H.; D'hooge, J. Multi-transmit beam forming for fast cardiac imaging-a simulation study. *IEEE Trans. Ultrason. Ferroelectr. Freq. Control* **2013**, *60*, 1719–1731.
5. Montaldo, G.; Tanter, M.; Bercoff, J.; Benech, N.; Fink, M. Coherent plane-wave compounding for very high frame rate ultrasonography and transient elastography. *IEEE Trans. Ultrason. Ferroelectr. Freq. Control* **2009**, *56*, 489–506.
6. Synnevag, J.; Austeng, A.; Holm, S. Adaptive beamforming applied to medical ultrasound imaging. *IEEE Trans. Ultrason. Ferroelectr. Freq. Control* **2007**, *54*, 1606.
7. Holfort, I.K.; Gran, F.; Jensen, J.A. Broadband minimum variance beamforming for ultrasound imaging. *IEEE Trans. Ultrason. Ferroelectr. Freq. Control* **2009**, *56*, 314–325.
8. Synnevag, J.F.; Austeng, A.; Holm, S. A low-complexity data-dependent beamformer. *IEEE Trans. Ultrason. Ferroelectr. Freq. Control* **2011**, *58*, 281–289.

9. Li, P.C.; Li, M.L. Adaptive imaging using the generalized coherence factor. *IEEE Trans. Ultrason. Ferroelectr. Freq. Control* **2003**, *50*, 128–141.
10. Camacho, J.; Parrilla, M.; Fritsch, C. Phase coherence imaging. *IEEE Trans. Ultrason. Ferroelectr. Freq. Control* **2009**, *56*, doi:10.1109/TUFFC.2009.1128.
11. Matrone, G.; Savoia, A.S.; Caliano, G.; Magenes, G. The delay multiply and sum beamforming algorithm in ultrasound B-mode medical imaging. *IEEE Trans. Med. Imaging* **2015**, *34*, 940–949.
12. Matrone, G.; Ramalli, A.; Savoia, A.S.; Tortoli, P.; Magenes, G. High frame-rate, high resolution ultrasound imaging with multi-line transmission and filtered-delay multiply and sum beamforming. *IEEE Trans. Med. Imaging* **2017**, *36*, 478–486.
13. Prieur, F.; Rindal, O.M.H.; Holm, S.; Austeng, A. Influence of the Delay-Multiply-And-Sum beamformer on the ultrasound image amplitude. In Proceedings of the 2017 IEEE International Ultrasonics Symposium (IUS), Washington, DC, USA, 6–9 September 2017.
14. Polichetti, M.; Varray, F.; Matrone, G.; Savoia, A.S.; Béra, J.C.; Cachard, C.; Nicolas, B. A computationally efficient nonlinear beamformer based on p-th root signal compression for enhanced ultrasound B-mode imaging. In Proceedings of the 2017 IEEE International Ultrasonics Symposium (IUS), Washington, DC, USA, 6–9 September 2017.
15. Liebgott, H.; Rodriguez-Molares, A.; Cervenansky, F.; Jensen, J.A.; Bernard, O. Plane-wave imaging challenge in medical ultrasound. In Proceedings of the 2016 IEEE International Ultrasonics Symposium (IUS), Tours, France, 18–21 September 2016.
16. Matrone, G.; Savoia, A.S.; Magenes, G. Filtered Delay Multiply And Sum beamforming in plane-wave ultrasound imaging: Tests on simulated and experimental data. In Proceedings of the 2016 IEEE International Ultrasonics Symposium (IUS), Tours, France, 18–21 September 2016.
17. Varray, F.; Bernard, O.; Assou, S.; Cachard, C.; Vray, D. Hybrid strategy to simulate 3D nonlinear radio-frequency ultrasound using a variant spatial PSF. *IIEEE Trans. Ultrason. Ferroelectr. Freq. Control* **2016**, *63*, 1390–1398.
18. Rindal, O.M.H.; Rodriguez-Molares, A.; Austeng, A. The dark region artifact in adaptive ultrasound beamforming. In Proceedings of the 2017 IEEE International Ultrasonics Symposium (IUS), Washington, DC, USA, 6–9 September 2017.
19. Thijssen, J.M. Ultrasonic speckle formation, analysis and processing applied to tissue characterization. *Pattern Recognit. Lett.* **2003**, *24*, 659–675.
20. Hverven, S.M.; Rindal, O.M.H.; Rodriguez-Molares, A.; Austeng, A. The influence of speckle statistics on contrast metrics in ultrasound imaging. In Proceedings of the 2017 IEEE International Ultrasonics Symposium (IUS), Washington, DC, USA, 6–9 September 2017.
21. Ramalli, A.; Scaringella, M.; Matrone, G.; Dallai, A.; Boni, E.; Savoia, A.S.; Bassi, L.; Hine, G.E.; Tortoli, P. High dynamic range ultrasound imaging with real-time filtered-delay multiply and sum beamforming. In Proceedings of the 2017 IEEE International Ultrasonics Symposium (IUS), Washington, DC, USA, 6–9 September 2017.
22. Errico, C.; Pierre, J.; Pezet, S.; Desailly, Y.; Lenkei, Z.; Couture, O.; Tanter, M. Ultrafast ultrasound localization microscopy for deep super-resolution vascular imaging. *Nature* **2015**, *527*, 499–502.
23. Boulos, P.; Varray, F.; Poizat, A.; Kalkhoran, M.A.; Gilles, B.; Bera, J.; Cachard, C. Passive cavitation imaging using different advanced beamforming methods. In Proceedings of the 2016 IEEE International Ultrasonics Symposium (IUS), Tours, France, 18–21 September 2016.

© 2018 by the authors. Licensee MDPI, Basel, Switzerland. This article is an open access article distributed under the terms and conditions of the Creative Commons Attribution (CC BY) license (http://creativecommons.org/licenses/by/4.0/).

Article

Sound Velocity Estimation and Beamform Correction by Simultaneous Multimodality Imaging with Ultrasound and Magnetic Resonance

Ken Inagaki, Shimpei Arai, Kengo Namekawa and Iwaki Akiyama *

Medical Ultrasound Research Center, Doshisha University, Kyotanabe Kyoto 610-0321, Japan;
ctub1016@mail4.doshisha.ac.jp (K.I.); ctub1002@mail4.doshisha.ac.jp (S.A.);
bmp1062@mail4.doshisha.ac.jp (K.N.)
* Correspondence: iakiyama@mail.doshisha.ac.jp

Received: 14 September 2018; Accepted: 27 October 2018; Published: 2 November 2018

Abstract: Since the sound velocity for medical ultrasound imaging is usually set at 1540 m/s, the ultrasound imaging of a patient with a thick layer of subcutaneous fat is degraded due to variations in the sound velocity. This study proposes a method of compensating for image degradation to correct beamforming. This method uses the sound velocity distribution measured in simultaneous ultrasound (US) and magnetic resonance (MR) imaging. Experiments involving simultaneous imaging of an abdominal phantom and a human neck were conducted to evaluate the feasibility of the proposed method using ultrasound imaging equipment and a 1.5 T MRI scanner. MR-visible fiducial markers were attached to an ultrasound probe that was developed for use in an MRI gantry. The sound velocity distribution was calculated based on the MRI cross section, which was estimated as a corresponding cross section of US imaging using the location of fiducial markers in MRI coordinates. The results of the abdominal phantom and neck imaging indicated that the estimated values of sound velocity distribution allowed beamform correction that yielded compensated images. The feasibility of the proposed method was then evaluated in terms of quantitative improvements in the spatial resolution and signal-to-noise ratio.

Keywords: beamforming; MRI; MR-visible fiducial marker; subcutaneous fat layer; thyroid imaging; spatial resolution; signal-to-noise ratio (SNR); 1-3 piezocomposite material

1. Introduction

Ultrasound diagnostic equipment, which has widely been used in clinical diagnosis, assumes that sound velocity has a uniform distribution, since it is based on pulse-echo imaging. However, in biological soft tissues, the sound velocity ranges from 1400 m/s to 1600 m/s [1]. For example, the sound velocity in mammalian fat tissues ranges from 1400 m/s to 1490 m/s [1]. Therefore, since the sound velocity for ultrasound imaging (US imaging) is usually assumed to be 1540 m/s, the imaging of a patient with a thick layer of subcutaneous fat is degraded due to variations in the sound velocity. Significant image degradation presumably reduces spatial resolution and the signal-to-noise ratio (SNR) in US imaging. If the sound velocity distribution is known before imaging, image degradation could be reduced by compensating for improper beamformation. Various methods for the in vivo measurement of the sound velocity distribution have been proposed [2–5]. For example, Aoki et al. [6] and Nitta et al. [7] proposed methods for the measurement of the sound velocity in cartilage using magnetic resonance imaging (MRI) and US imaging. In the two methods, sound velocity is estimated based on the length of cartilage, as measured with MRI, divided by the time-of-flight, as measured with the ultrasonic pulse-echo technique. Since the estimated sound velocity is obtained at different times and locations in US imaging and MRI, the measurements are not accurate enough to correct beamforming.

Thus, the aim of the current study was to evaluate the feasibility of a method of measuring the sound velocity distribution in vivo using a prototype simultaneous US imaging and MRI system. This method corrected beamforming to improve image quality in terms of the spatial resolution and SNR. In this study, magnetic-resonance-visible (MR-visible) fiducial markers were attached to an ultrasound probe to display a cross section of US imaging in MRI coordinates. The ultrasound probe was developed for use in the coils of an MRI scanner.

When measuring the sound velocity distribution with a simultaneous US imaging and MRI system, the following issues need to be resolved: (1) the development of an ultrasonic array probe for use in an MRI gantry; (2) the development of MR-visible fiducial markers and the estimation of the location of a cross section in MRI coordinates; and (3) the suppression of crosstalk (electrical noise) between the MRI scanner and ultrasound equipment. Curiel et al. [8] imaged a phantom kidney and a rabbit kidney by mechanically scanning a single transducer. Tang et al. [9] used a commercially available probe with attached MR-visible fiducial markers to simultaneously image a phantom. In the current study, linear array transducers made of 1-3 piezocomposite materials [10,11] were embedded in an ultrasound probe for use in an MRI gantry. The probe was made of nonmagnetic materials, and MR-visible fiducial markers were attached to it. The probe used in the gantry was connected to ultrasound equipment via a connector that passed through the walls shielding the MRI room and the control room. Crosstalk between the ultrasound equipment and the MRI scanner was reduced by grounding the connector.

Simultaneous imaging of an abdominal phantom and a human neck was performed to evaluate the feasibility of the proposed method. This was accomplished with US imaging equipment with 128 transmission/reception channels and a 1.5 T MRI scanner. Imaging of the phantom indicated that the error rate for the accuracy of sound velocity measurement was less than 6%, while the spatial resolution was 0.43 (the ratio of the lateral resolution before and after compensation), and the SNR was 8 dB. Simultaneous imaging of a human neck was performed to evaluate the compensation for image degradation due to the layer of subcutaneous fat. An acoustic standoff pad with a sound velocity similar to that of the fat tissue was inserted between the probe and the surface of the skin on the neck. Improvement in image quality was noted in the resulting image of the thyroid region. During imaging of the thyroid region in the neck, the spatial resolution (the ratio of the lateral resolution before and after compensation) was 0.60 and the SNR improved by 3 dB as compared to the SNR in the Region of Interest (ROI) without compensation. Nevertheless, the clinical feasibility of the proposed method, and especially its use in abdominal imaging, needs to be studied further.

2. Materials and Methods

2.1. Ultrasound Probe for Use in MRI

An ultrasound probe for use in an MRI gantry was developed to evaluate the feasibility of simultaneous US imaging and MRI. Array transducers made of 1-3 piezocomposite materials were embedded in the probe (Kyokutan, Japan Probe Co., Yokohama, Japan) as shown in Figure 1. The specifications of the array transducers are shown in Table 1. MR-visible fiducial markers were attached to the probe to indicate the probe position and orientation in a three-dimensional MR image. Each MR-visible fiducial marker was made of a polyoxymethylene (POM) sphere contained in an acrylic cylinder (9.00 mm in diameter and 8.50 mm in height) filled with olive oil, as shown in Figure 2. Since the marker appeared in an MRI as a dark sphere within a bright cylinder, the geometric center of the dark region served as the marker location. The eight markers were arranged in two rows in the probe as shown in Figure 1. The array transducers were arranged along the center line between the two rows. Thus, the location of the cross section of the ultrasound image was estimated based on the coordinates of the marker location in the MRI image. The transducer array was located along the central line between two rows formed by MR-visible fiducial markers. Thus, the cross section of an MRI image corresponding to a cross section of an ultrasound image was estimated as the plane through the central line and perpendicular to the lines formed by the pairs of markers. The central coordinates of each marker were calculated by MRI. The distance between the footprint of the transducer array and the marker array was measured in advance. Thus, the cross section of the ultrasound B-scan image was located in the coordinates of the MRI.

Table 1. Specifications of the array transducers used in MRI.

Bandwidth (MHz)	Number of Elements	Element Pitch (mm)	Element Size (mm)	Focal Length of Acoustic Lens (mm)
5–8	192	0.30	8.0 × 0.20	20

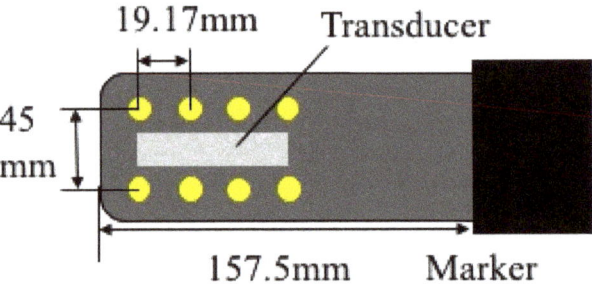

Figure 1. Top view of the prototype ultrasound probe for use in MRI. Array transducers made of 1-3 piezocomposite materials were embedded in the probe (Kyokutan, Japan Probe Co., Yokohama, Japan). The specifications of the array transducers are shown in Table 1. MR-visible fiducial markers were attached to the probe to locate the probe position and orientation in a three-dimensional MRI image.

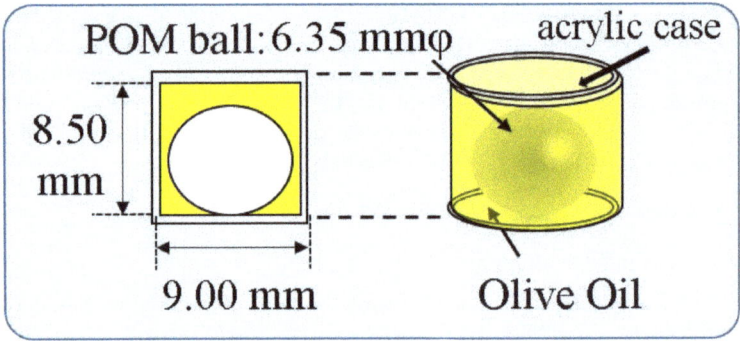

Figure 2. MR-visible fiducial marker. (POM: polyoxymethylene).

2.2. Simultaneous Multimodality Imaging System Using Ultrasound and Magnetic Resonance

This study used a simultaneous multimodality imaging system consisting of an MRI scanner (Echelon Vega, 1.5T, Hitachi Co., Tokyo, Japan) and US imaging equipment (RSYS0006MRF, Microsonic Co., Tokyo, Japan). The US imaging equipment was located in the control room as shown in Figure 3. The specifications of the US imaging equipment are shown in Table 2. The probe that was placed in the MRI scanner was connected to the US imaging equipment via a connector that passed through the walls shielding the MRI room and the control room. This connector was grounded in the MRI room to eliminate crosstalk between the MRI scanner and the US imaging equipment.

A B-scan image was formed by a dynamic focusing method applied to the echo signals received by array transducers. This focusing was obtained by the summation of delayed ultrasonic echo signals.

The delay time sequence was calculated by the sound velocity and the path between an element of the array and a virtual focal point. Due to the conventional focusing method used, the sound velocity was assumed to have a constant value of 1540 m/s. The beamforming method proposed in this paper is a dynamic focusing technique that employs the sound velocity and the path estimated by simultaneous US imaging and MRI. This method can also be applied to the transmitting beam focusing technique if the sound velocity distribution is obtained by MRI. Through the use of simultaneous US imaging and MRI, beamform correction can be achieved for both the transmission and reception. Therefore, this method offers a novel approach to beamform correction.

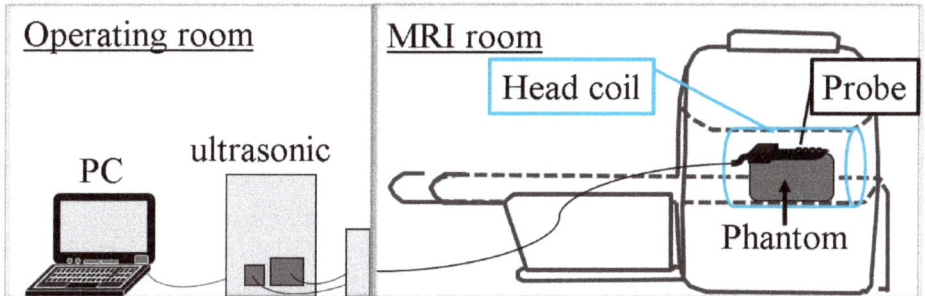

Figure 3. A simultaneous multimodality imaging system consisting of ultrasound (US) imaging equipment (RSYS0006MRF, Microsonic Co., Tokyo, Japan) and an MRI scanner (Echelon Vega, 1.5T, Hitachi Co., Tokyo, Japan). The probe that was placed in the MRI scanner was connected to the US imaging equipment via a connector that passed through the walls shielding the MRI room and the control room.

Table 2. Specifications of US imaging equipment.

Maximum Number of Probe Interface Channels	Number of TX/RX Channels	A/D Resolution (Bits)	Sampling Frequency (Hz)	Memory Capacitance Channel (MB)
256	128	12	31.25	256

2.3. Measurement Method of Sound Velocity in Human Crus

An experiment was conducted to measure the sound velocity in human fat and muscle in the crus using the method of simultaneous US imaging and MRI. The probe was placed on the top of the crus. The focal length for the transmission beam was set at 30 mm. The MRI scan parameters for human crus imaging are shown in Table 3. An RF receiving coil for the knee was used in the MRI setup. The thickness, L, of muscle and fat was measured by MRI. The time-of-flight, T, was measured by using an ultrasonic echo signal. Thus, the sound velocity, c, was estimated to be c = 2L/T. The crura of three subjects were measured.

Table 3. MRI scan parameters for human crus.

Sequence	TR	TE	Thickness	Slices	Frequency Encoding	Phase Encoding
SE	937 ms	19.7 ms	2.0 mm	30	256	180

2.4. Experimental Imaging of an Abdominal Phantom using the Proposed Method of Compensating for the Sound Velocity Distribution to Correct Beamforming

An experiment was conducted to evaluate the accuracy with which the sound velocity distribution was measured during simultaneous US imaging and MRI of an abdominal phantom. Image degradation was compensated for by correcting beamforming using the measured sound velocity distribution. The Model 075A abdominal phantom (Triple Modality 3D Abdominal Phantom, CIRS, USA) was used in the experiment. The probe was placed on the top of the phantom. Agar gel containing 5% glycerol was placed between the phantom and the probe to avoid air cavities, as shown in Figure 4. The driving waveform to the transducers was a square wave containing 1.5 sinusoidal waves. The pulse repetition time was 190 µs. The focal length for the transmission beam was set at 30 mm using the conventional array focusing technique. Since the phantom consists of four regions with different sound velocities as shown in Figure 4, the sound velocity in each region was estimated using simultaneous US imaging and MRI. For each region, the length was represented by L and the time-of-flight was represented by T, so the sound velocity c was estimated to be c = 2L/T. The delay time for each element to form a beam must be calculated based on the estimated value of sound velocity on the path from an element to an arbitrary position during simultaneous imaging.

The signal processing was as follows: (1) RF data acquisition of 128 elements in a 128-channel digital format at 12 bits and 31.25 MHz; (2) the RF data of each channel were delayed by a certain amount of time calculated by Equations (1) or (2); (3) after quadrature detection at 5.0 MHz, the magnitude was calculated; and (4) the B-scan image was calculated using the logarithmic magnitude in a dynamic range of 60 dB.

For the conventional dynamic focusing in reception, the delay time of the *i*th element of the transducers τ_i is expressed as

$$\tau_i = -\frac{(i\Delta x - a/2)^2}{2cl_F} \quad (1)$$

where Δx is the element pitch, a is the aperture length, c is the sound velocity, and l_F is the focal length.

For the proposed compensated dynamic focusing in reception, the delay time of the *i*th element of the transducers τ_i is expressed as

$$\tau_i = -\sum_{k=1}^{N} \frac{l_k}{c_k} + \sum_{k=1}^{N_0} \frac{l_{0k}}{c_{0k}} \quad (2)$$

where N is the number of the region with the same sound velocity along the ultrasound propagation path from the *i*th element to the focal point, c_k is the sound velocity in the *k*th region, and l_k is the length of the path through the *k*th region. The subscript of 0 indicates the path from the center element of the aperture to the focal point. The region with the same sound velocity is estimated by MRI. MRI scan parameters for abdominal phantom imaging are shown in Table 4.

Table 4. MRI scan parameters for the abdominal phantom and human neck.

Sequence	TR	TE	Thickness	Slices	Frequency Encoding	Phase Encoding
SE	1436 ms	20 ms	1.5 mm	45	256	180

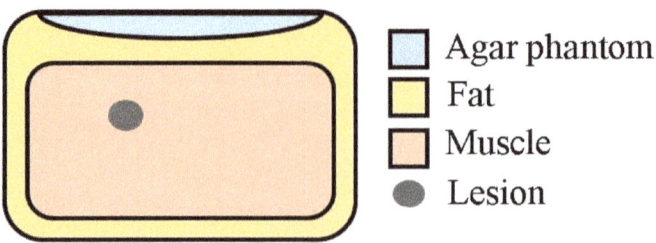

Figure 4. Composition of the phantom used in the experiment.

2.5. Experimental Imaging of the Human Neck

An experiment was conducted to measure the sound velocity distribution during simultaneous US imaging and MRI of the human neck. Image degradation due to a layer of subcutaneous fat was compensated for by correcting beamforming. The experimental setup was the same as that shown in Figure 3. Image quality significantly degrades due to a layer of subcutaneous fat in the abdominal region. An MRI scan takes 20 min. As the imaging time increases, the spatial resolution decreases due to heartbeat and respiration. MRI images need to be obtained with a high spatial resolution in order to significantly improve image quality. Thus, the neck region of a male healthy volunteer was imaged to avoid motion artifacts. Since the subject had a thin layer of subcutaneous fat in his neck, an acoustic standoff pad (Model AC-1, Acoustic standoff pads, ATS Laboratories Co., Nofolk, VA, USA) was placed between the ultrasound probe and the surface of the skin of the neck to mimic a layer of subcutaneous fat. This pad had a thickness of 10 mm and a sound velocity of 1410 m/s [12]. MRI scan parameters for human neck imaging are shown in Table 4.

3. Results and Discussion

3.1. Experimental Results of the Measurement of Sound Velocity in Human Crus

The transducer array was located along the central line between two rows formed by MR-visible fiducial markers. Thus, the cross section of an MRI image corresponding to a cross section of an ultrasound image was estimated as the plane through the central line and perpendicular to the lines formed by the pairs of markers. The resultant cross section of the MRI image is shown in Figure 5.

The estimated values of the sound velocity in human muscle and fat tissue in the crus for three subjects are shown in Table 5. The sound velocities in fat and muscle were measured as 1450 ± 20 m/s and 1560 ± 20 m/s, respectively. The sound velocities in human fat and muscle measured in vitro are typically in the ranges of 1459–1479 m/s and 1540–1566 m/s, respectively [1]. Those values are in approximate agreement with those obtained in this study, i.e., falling within the standard deviation.

Figure 5. An example of an MRI image of a human crus. The fat layer corresponds to the upper high-brightness region. The sound velocity in fat was measured in the central fat layer. The sound velocity in muscle was measured in the central muscle region between the bottom of the fat layer and the top of the fascia layer (lower-brightness horizontal stripe region).

Table 5. The estimated values of sound velocity obtained by simultaneous US imaging and MRI.

		Subject 1	Subject 2	Subject 3	Average	References [1]
Fat layer	Length (mm)	3.58 ± 0.1	4.63 ± 0.1	4.24 ± 0.1	-	-
	Time of flight (µs)	5.02 ± 0.03	6.34 ± 0.03	5.86 ± 0.03	-	-
	Sound velocity (m/s)	1430 ± 40	1460 ± 30	1450 ± 40	1445 ± 20	1459–1479
Muscle tissue	Length (mm)	15.00 ± 0.1	14.13 ± 0.1	15.7 ± 0.1	-	-
	Time of flight (µs)	9.75 ± 0.03	17.92 ± 0.03	20.0 ± 0.03	-	-
	Sound Velocity (m/s)	1540 ± 10	1580 ± 10	1570 ± 10	1562 ± 30	1540–1566

3.2. Imaging of an Abdominal Phantom

The results of the simultaneous imaging of an abdominal phantom are described as follows. The cross section of an MRI image was estimated based on MR-visible fiducial markers in the corresponding cross section of an ultrasound image, as shown in Figure 6. The sound velocity was estimated to be 1500 m/s in the agar gel and 1430 m/s in the fat-mimicking region. Image degradation was compensated for by using a sound velocity of 1540 m/s for the soft-tissue-mimicking region. The resulting image was compared to an ultrasound image that was obtained by assuming a sound velocity with a uniform distribution as shown in Figure 7. Fine specks are evident in the lower layer of fat in the compensated image, as shown in Figure 7a, although the margins of darker regions are indistinct. Similar specks are evident in the horizontal direction in a conventional B-scan image as shown in Figure 7b. This image was obtained when a constant sound velocity of 1540 m/s was used in beamforming. The margins of the darker circular region are clearly evident compared to those in the compensated image. Image degradation due to a fluctuation in sound velocity was improved by correcting beamforming using the estimated sound velocity distribution. To improve the spatial resolution, the half-width of the autocorrelation function in the horizontal direction was calculated for the three red squares shown in Figure 7; the half-width of the autocorrelation function is defined as the distance equal to half of the maximum value of the autocorrelation function. The calculated half-widths are shown in Table 6. The average ratio of improvement was 0.43. The SNR is defined as the signal power divided by the noise power; the signal power is calculated as the square of the maximum value in the transmitting focal region and the noise power is calculated as the standard deviation in the ROI. The improvement in the SNR was calculated to be 8 dB based on the images shown in Figure 7a,b.

Figure 6. Cross section of an MRI image corresponding to that of an ultrasound image.

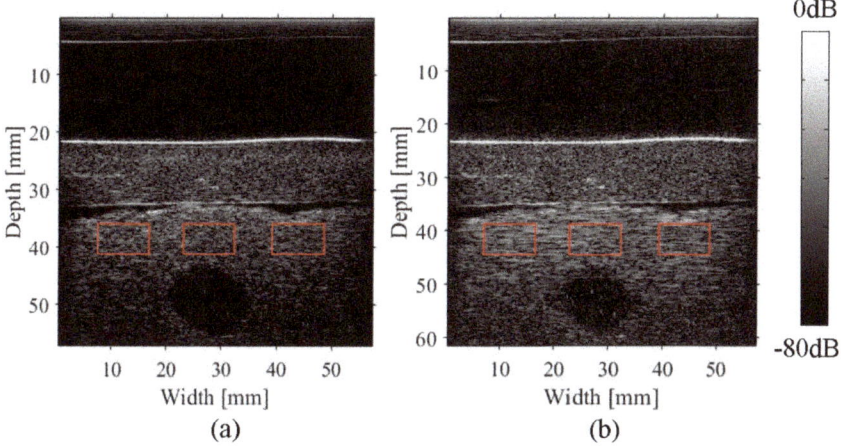

Figure 7. Comparison of B-mode images of the phantom obtained using the proposed method of compensation (**a**) and a conventional approach (**b**). The three red squares indicate ROIs for calculation of the autocorrelation function for a B-mode image.

Table 6. Comparison of the half-widths of the autocorrelation functions in the lateral direction.

ROIs	The Half-Width of the Autocorrelation Function for a Compensated Image	The Half-Width of the Autocorrelation Function for a Conventional Image	Ratio of the Half-Width Improvement
Left square	1.35 mm	2.55 mm	0.53
Center square	0.85 mm	2.60 mm	0.30
Right square	1.33 mm	3.02 mm	0.44
Average	1.18 mm	2.72 mm	0.43

3.3. Imaging of the Neck

The simultaneous application of US imaging and MRI to the neck was performed to determine the feasibility of the proposed method of compensation. Examples of T1-weighted images that were obtained with the simultaneous imaging system are shown in Figure 8. The MR-visible fiducial markers that were attached to the ultrasound probe are indicated by the arrows in Figure 8. The cross section of an MR image corresponding to the cross section of an ultrasound image is shown in Figure 9. First, the sound velocity in the acoustic gel pad was estimated as 1360 m/s by simultaneous US imaging and MRI. The difference between the estimated value and the reference value was 50 m/s. Beamforming was corrected using an estimated sound velocity of 1360 m/s for the pad and a sound velocity of 1540 m/s for biological tissues in the neck region. The resulting image was compared to a conventional B-scan image as shown in Figure 10. Image degradation due to variations in the sound velocity was reduced by compensating for the sound velocity, as shown in Figure 10a. Without compensation, the conventional B-scan image was blurred as a result of fluctuations in the sound velocity, as shown in Figure 10b. To determine the improvement in the spatial resolution, the half-width of the autocorrelation function in the horizontal direction was calculated for the region of the red square shown in Figure 10. The ratio of the half-width of the autocorrelation in the conventional image with that in the compensated image was calculated to be 0.60. The improvement in the SNR was 3 dB. The results indicated that simultaneous US imaging and MRI is a feasible way to reduce image degradation due to layers of subcutaneous fat by compensating for the sound velocity to correct beamforming.

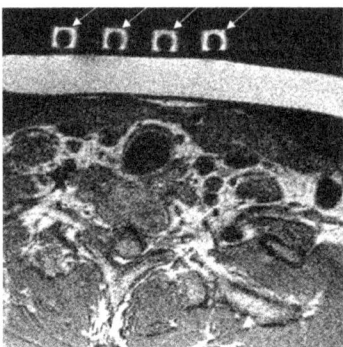 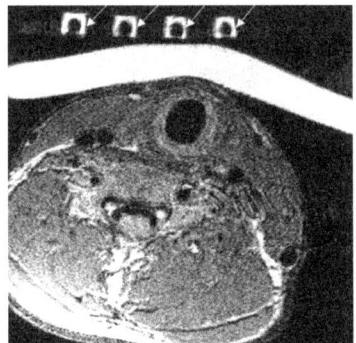

Figure 8. Examples of MR images obtained using the simultaneous multimodality imaging system. Each image corresponds to a cross section, and rows of fiducial marker arrays are indicated by arrows. The probe was equipped with two rows of marker arrays, as shown in Figure 1.

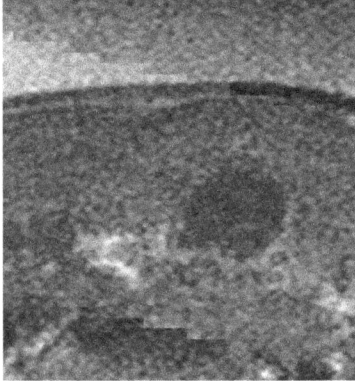

Figure 9. The estimated cross section of the MR image corresponding to the cross section of the ultrasound image according to coordinates of MR-visible fiducial markers.

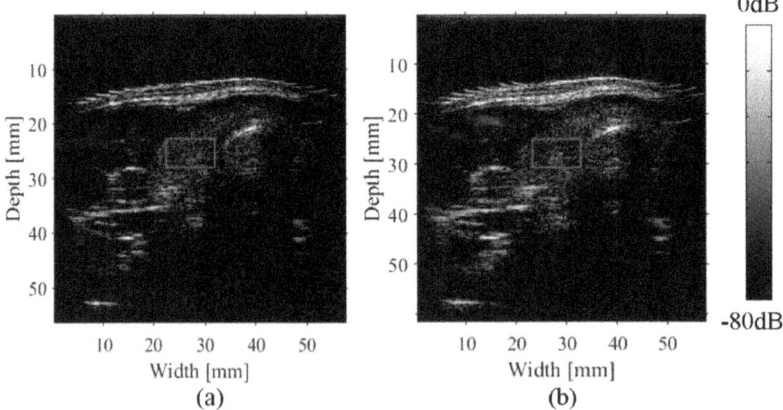

Figure 10. B-mode images of the neck obtained using the proposed method of compensation (**a**) and a conventional approach (**b**). A red square indicates a ROI in part of the thyroid for the autocorrelation function in each B-mode image.

4. Conclusions

US imaging based on the pulse-echo method assumes that the sound velocity has a constant value in the imaging media. In general, a conventional B-scan imaging system uses a sound velocity of 1540 m/s as the mean value for biological soft tissues. Since the sound velocity for biological tissues ranges from 1400 m/s to 1600 m/s, differences in the sound velocity affect beamforming and cause image degradation. In particular, the sound velocity for fat tissue is 10% slower than the mean value for biological soft tissues, so a thick layer of subcutaneous fat degrades image quality in terms of the spatial resolution and SNR. Thus, this study proposed a method of compensating for image degradation due to fluctuations in sound velocity. This method uses simultaneous multimodality imaging with US and magnetic resonance to estimate the distribution of the sound velocity in order to correct beamforming. An experiment was conducted with a phantom and imaging of a human neck was performed to evaluate the feasibility of the proposed method. A cross section of an MRI image and the corresponding cross section of an ultrasound image were compared to accurately measure the sound velocity. An MR-visible fiducial marker was developed and attached to an ultrasound probe for use in an MRI gantry. The experiment with a phantom indicated that the spatial resolution and SNR improved. During imaging of the human neck, an acoustic standoff pad was placed between the ultrasound probe and the surface of the skin of the neck to mimic a layer of subcutaneous fat. The sound velocity in the pad was estimated using simultaneous US imaging and MRI. Imaging of the neck indicated that the correction of the beamforming using the estimated sound velocity resulted in improved image quality for the thyroid region. The current results indicated that the proposed method is a feasible form of beamform correction in terms of improvements in the spatial resolution and SNR.

Author Contributions: Conceptualization, I.A.; methodology, K.I., I.A.; software, K.I.; validation, K.I., I.A.; formal analysis, K.I., I.A.; investigation, K.I., S.A., K.N.; resources, I.A.; data curation, K.I., S.A., K.N.; writing—original draft preparation, K.I.; writing—review and editing, K.I., I.A.; visualization, K.I., S.A.; supervision, I.A.; project administration, I.A.; funding acquisition, I.A.

Acknowledgments: This study was supported by the MEXT-Support Program to Strategic Research Foundation at Private Universities, 2013–2017. The authors wish to thank Kazuya Niyagawa of Ricoh Co., Ltd. for his help with fiducial processing.

Conflicts of Interest: The authors declare no conflict of interest.

References

1. Goss, S.A.; Johnston, R.L.; Dunn, F. Comprehensive compilation of empirical ultrasonic properties of mammalian tissues. *J. Acoust. Soc. Am.* **1978**, *64*, 423–457. [CrossRef] [PubMed]
2. Robinson, D.E.; Ophir, J.; Wilson, L.S.; Chen, C.F. Pulse-Echo Ultrasound Speed Measurements: Progress and Prospects. *Ultrasound Med. Biol.* **1991**, *17*, 633–646. [CrossRef]
3. Kondo, M.; Takamizawa, K.; Hirama, M.; Okazaki, K.; Iinuma, K.; Takehara, Y. An evaluation of an in vivo local sound speed estimation technique by the crossed beam method. *Ultrasound Med. Biol.* **1990**, *16*, 65–72. [CrossRef]
4. Anderson, M.E.; Trahey, G.E. The direct estimation of sound speed using pulse–echo ultrasound. *J. Acoust. Soc. Am.* **1998**, *104*, 3099–3106. [CrossRef] [PubMed]
5. Benjamin, A.; Zubajlo, R.E.; Dhyani, M.; Samir, A.E.; Thomenius, K.E.; Grajo, J.R.; Anthony, B.W. Surgery for Obesity and Related Diseases: I. A Novel Approach to the Quantification of The Longitudinal Speed of Sound and Its Potential for Tissue Characterization. *Ultrasound Med. Biol.* **2018**. [CrossRef] [PubMed]
6. Aoki, T.; Nitta, N.; Furukawa, A. Non-invasive speed of sound measurement in cartilage by use of combined magnetic resonance imaging and ultrasound: An initial study. *Radiol. Phys. Technol.* **2013**, *6*, 480–485. [CrossRef] [PubMed]
7. Nitta, N.; Kaya, A.; Misawa, M.; Hyodo, K.; Numano, T. Accuracy improvement of multimodal measurement of speed of sound based on image processing. *Jpn. J. Appl. Phys.* **2017**, *56*, 07JF17. [CrossRef]
8. Curiel, L.; Chopra, R.; Hynynen, K. Progress in Multimodality Imaging: Truly Simultaneous Ultrasound and Magnetic Resonance Imaging. *IEEE Trans. Med. Imaging* **2007**, *26*, 1740–1746. [CrossRef] [PubMed]

9. Tang, A.M.; Kacher, D.F.; Lam, E.Y.; Wong, K.K.; Jolesz, F.A.; Yang, E.S. Simultaneous Ultrasound and MRI System for Breast Biopsy: Compatibility Assessment and Demonstration in a Dual Modality Phantom. *IEEE Trans. Med. Imaging* **2008**, *27*, 247–254. [CrossRef] [PubMed]
10. Smith, W.A.; Shaulov, A.; Auld, B.A. Tailoring the properties of composite piezoelectric materials for medical ultrasonic transducers. In Proceedings of the 1985 IEEE Ultrasonics Symposium, San Francisco, CA, USA, 16–18 October 1985; pp. 642–647.
11. Smith, W.A. Modeling 1–3 Composite Piezoelectrics: Hydrostatic Response. *IEEE Trans. Ultrason. Ferroelectr. Freq. Control* **1993**, *40*, 41–49. [CrossRef] [PubMed]
12. Lynch, T.; (Computerized Imaging Reference Systems, Incorporated (CIRS)). Personal communication, 2018.

© 2018 by the authors. Licensee MDPI, Basel, Switzerland. This article is an open access article distributed under the terms and conditions of the Creative Commons Attribution (CC BY) license (http://creativecommons.org/licenses/by/4.0/).

Article

Multi-Perspective Ultrasound Imaging Technology of the Breast with Cylindrical Motion of Linear Arrays

Chang Liu [1,2], Binzhen Zhang [1,*], Chenyang Xue [1,*], Wendong Zhang [1], Guojun Zhang [1] and Yijun Cheng [3]

[1] Key Laboratory of Instrumentation Science & Dynamic Measurement, North University of China, Ministry of Education, Taiyuan 030051, China; liuchang870820@126.com (C.L.); wdzhang@nuc.edu.cn (W.Z.); zhangguojun1977@nuc.edu.cn (G.Z.)
[2] School of Electrical and Electronic Engineering, Dalian Vocational Technical College, Dalian 116037, China
[3] Department of Electronic and Engineering, Taiyuan Institute of Technology, Taiyuan 030008, China; alexcheng@163.com
* Correspondence: zhangbinzhen@nuc.edu.cn (B.Z.); xuechenyang@nuc.edu.cn (C.X.); Tel.: +86-351-3921-756 (B.Z.); +86-351-3921-882 (C.X.)

Received: 8 January 2019; Accepted: 23 January 2019; Published: 26 January 2019

Abstract: In this paper, we propose a multi-perspective ultrasound imaging technology with the cylindrical motion of four piezoelectric micromachined ultrasonic transducer (PMUT) rotatable linear arrays. The transducer is configured in a cross shape vertically on the circle with the length of the arrays parallel to the z axis, roughly perpendicular to the chest wall. The transducers surrounded the breast, which achieves non-invasive detection. The electric rotary table drives the PMUT to perform cylindrical scanning. A breast model with a 2 cm mass in the center and six 1-cm superficial masses were used for the experimental analysis. The detection was carried out in a water tank and the working temperature was constant at 32 °C. The breast volume data were acquired by rotating the probe 90° with a 2° interval, which were 256 × 180 A-scan lines. The optimized segmented dynamic focusing technology was used to improve the image quality and data reconstruction was performed. A total of 256 A-scan lines at a constant angle were recombined and 180 A-scan lines were recombined according to the nth element as a dataset, respectively. Combined with ultrasound imaging algorithms, multi-perspective ultrasound imaging was realized including vertical slices, horizontal slices and 3D imaging. The seven masses were detected and the absolute error of the size was approximately 1 mm where even the image of the injection pinhole could be seen. Furthermore, the breast boundary could be seen clearly from the chest wall to the nipple, so the location of the masses was easier to confirm. Therefore, the validity and feasibility of the data reconstruction method and imaging algorithm were verified. It will be beneficial for doctors to be able to comprehensively observe the pathological tissue.

Keywords: multi-perspective ultrasound imaging; cylindrical scanning; dynamic focusing; PMUT linear array

1. Introduction

Breast cancer is one of the most harmful diseases to women in today's society. Therefore, early detection and early treatment of breast cancer is particularly crucial [1]. The breast is made up of breast glands and other soft tissues without bones, which the cancerous tissues have higher density [2,3]. In order to avoid the harm of false positive to women, it is necessary to develop highly sensitive and specific diagnostic tools for early detection of breast cancer [4]. At present, the imaging technology for breast cancer mainly includes X-ray mammography, Magnetic resonance imaging (MRI) and ultrasound [5]. X-ray mammography has the advantages of high sensitivity and high specificity to calcification points [6]. Whereas, early lesion is small, similar to the density of surrounding glands and the boundary is not clear, hence the imaging effect in breast detection is not obvious [7]. Besides, because of the radiation hazards, early screening is not appropriate [8]. Although MRI has higher sensitivity, the specificity is rather poor [9]. In addition, its disadvantages are high costs and long inspection time, which is not conducive to early screening [10–12]. Likewise, breast ultrasound relies heavily on the doctor's interpretation and is often used as a supplementary tool [13,14]. In general, biopsy is the gold standard for the diagnosis of breast cancer, which show that there are a lot of false-positive cases diagnosed by ultrasound imaging [15–18]. Ultrasound computer tomography (USCT) is a potential candidate for the imaging of breast cancer [19,20]. The simultaneous recording of reflection, speed of sound and attenuation images is the key superiority of the USCT system [21]. Comfort, safety and 3D imaging are the potential clinical benefits of ultrasound tomography [22,23].

Three-dimensional USCT, which is a new ultrasound imaging technology, promises high quality images with satisfactory reproducibility and could offer a better chance for survival by the detection of cancers at the early stage [24,25]. One problem that exists with the image reconstruction process is the assumption of 2D geometry while the object imaged is in 3D [26]. In a 2D image sequence, only clinicians with experience can estimate the size and shape of lesions and construct a three-dimensional geometric relationship between the lesion and its surrounding tissue [27]. This brings great difficulty to the accuracy and convenience of diagnosis and treatment and the specificity is poor [28]. Therefore, it is very necessary to visually demonstrate the collection data of USCT on a computer with different perspective [29]. Breast data were obtained from different perspectives and then ultrasound imaging analysis was conducted to observe tissue lesion information retrospectively, so as to improve specificity, further reduce misdiagnosis rate and avoid the pain caused by biopsy.

In this study, we addressed the multi-perspective imaging technology, which is used to diagnose breast lesions. A system with a cylindrical motion of four 1 × 128 PMUT linear arrays was applied to acquire the whole volume data using the optimized segmented dynamic focusing technology in the reflection mode. The data reconstruction was performed at a constant angle and according to the Nth element respectively. The experiment platform was set up and an ultrasound tomography algorithm was used to generate the vertical slice, horizontal slice and three-dimensional imaging. Furthermore, the characteristics of each perspective image were compared and analyzed.

2. Experiments and Methods

2.1. 3D Imaging System with Cylindrical Motion of Linear Arrays

We developed an original breast ultrasound imaging system with the cylindrical motion of four linear arrays. The setup was mainly used to diagnose breast lesions. Figure 1 shows the schematic diagram of the experimental setup and the connection of main components. Ultrasound imaging with the cylindrical motion of linear arrays is as shown in Figure 2. The transducer configuration scheme is shown in Figure 2a. The 1 × 128 PMUT was placed 90° cross vertically on the circle with the length of the arrays parallel to the axis of the circle, which was roughly parallel to the central axis of the breast. Figure 2b shows the dimensions (7 cm × 3 cm × 13.5 cm) of the linear array transducer and the active area (1.8 cm × 12.8 cm) of the transducer for imaging. The sensors were customized for the breast detection. The transducers surrounded the breast, which allowed for non-invasive detection. The detection was carried out in the water tank and the working temperature was constant at 32 °C. By controlling the electric rotary table controller through the PC workstation, the electric rotary table drives the PMUT to perform the cylindrical scanning. The cylindrical scanning diagram is shown in Figure 2c. The four PMUT linear arrays were arranged in a cross shape. The distance of the opposite PMUT was 180 mm and the array element center to center spacing was 1 mm. Hence, the maximum test aperture was 18 cm × 13 cm. Set 1 (Adjacent PMUT) and Set 2 (Another adjacent PMUT) operate alternately. A 64 channel ultrasonic transmitting/receiving acquisition circuit was employed to control the linear array. The data acquisition circuits had a sampling frequency of 40 MHz.

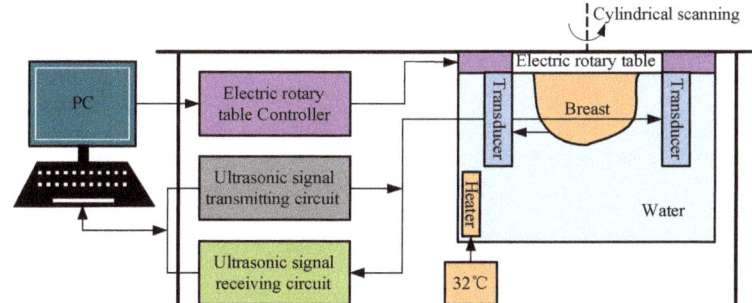

Figure 1. Schematic diagram of the experimental setup.

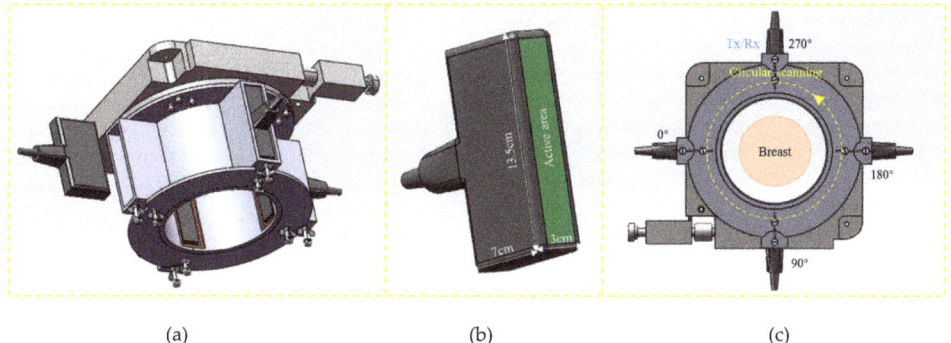

Figure 2. Ultrasound imaging with cylindrical motion of linear arrays. (**a**) Transducer configuration scheme. (**b**) The dimensions of the linear array transducer. (**c**) Cylindrical scanning diagram.

The breast volume data were obtained by performing the cylindrical scanning with a constant interval. For four linear array transducers, a full circle requires rotating the sensors 90 degrees. The step angle was set to 2°, therefore, it took 1.2 s for each rotation and 3 s for the vertical slice data collection. Hence, the whole scan time was about 3 minutes using a HP workstation (16G). The smaller the interval angle, the longer the scan time. Besides, the rotation precision was 0.05°, which allows for obtaining 7200 pulse echo signals. The optimized segmented dynamic focusing technology was used to improve the emission energy of the A-scan line, thereby improving the imaging quality. The high density was 256 scan lines and the low density was 128 scan lines, respectively; hence, the maximum data were 7200 × 256 A-scan lines. Multi-perspective ultrasound imaging of the breast could be obtained through the data reconstruction where the data could be reconstructed to obtain 7200 vertical slices, 256 horizontal slices and a three-dimensional imaging. By adopting four rotatable linear arrays to perform cylindrical scanning, it could make the sensor's fabrication process easier and gain better consistency and reliability than a cylindrical array [30].

The 1 × 128 PMUT linear array was characterized using a precision impedance analyzer (Agilent E4990A) as shown in Figure 3. The ultrasound transducer had a static capacitance of 664 pF, an impedance of 64 Ω and the operating frequency of 3.5 MHz [30]. Eight elements were randomly selected to test the transmitting sensitivity Sv and receiving sensitivity M as shown in Figure 3a. The average value of Sv is -220.5875 dB and the average value of M is 166.775 dB. The consistency is ±1dB@3.5MHz. The bandwidth experiment was conducted with the ultrasound transducer working as a transmitter and a standard transducer working as a receiver. The frequency response had a −6 dB bandwidth of 86.7% as seen in Figure 3b.

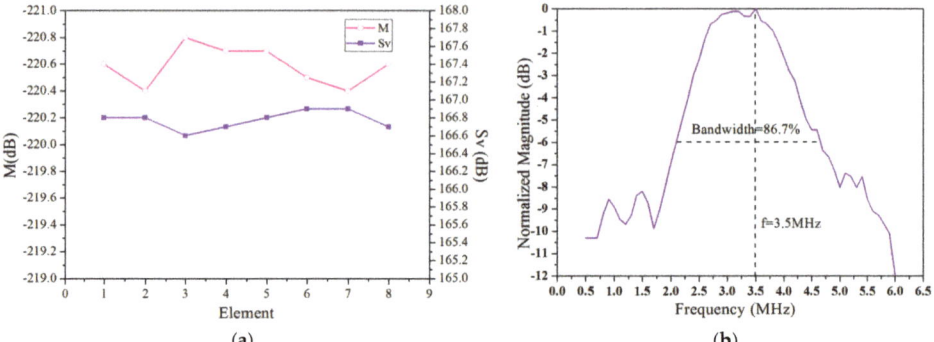

Figure 3. Frequency property of the ultrasound transducer. (**a**) Sensitivity. (**b**) Fractional bandwidth.

2.2. Multi-Perspective Ultrasound Imaging

This system was able to gain sufficient data and store it in the computer by rotating a full circle. These data could then be reconstructed to form multi-perspective breast ultrasound imaging as shown in Figure 4. Ultrasonic reflection imaging is mainly dependent on the acoustic impedance mismatch between different tissues [23]. In this paper, we describe the recombination method and ultrasound tomography algorithm from three perspectives, the details are as follows:

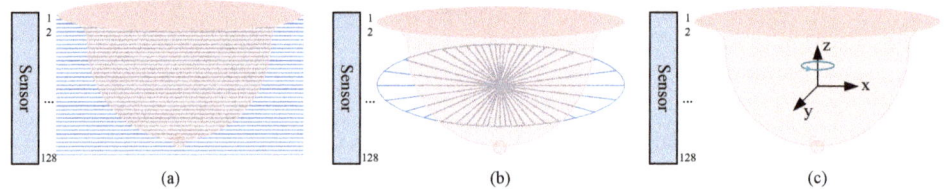

Figure 4. Multi-perspective breast ultrasound imaging. (**a**) Vertical slice (**b**) Horizontal slice. (**c**) 3D imaging.

2.2.1. Vertical Slices

A total of 256 A-scan lines at a constant angle were recombined as a dataset; then, we used a Butterworth filter to process the data for decreasing signal noise. Envelope detection was carried out for the data after filtering and the contour of the signal was extracted. The enveloping signal was processed by logarithmic compression and the dynamic range of the image was adjustable. Then, a vertical slice could be produced using gray-scale imaging as shown in Figure 4a.

2.2.2. Horizontal Slices

A total of 7200 A-scan lines were recombined according to the nth element as a dataset. Then, a Butterworth filter was used to process the data for decreasing signal noise. Envelope detection was carried out for the data after filtering and the contour of the signal was extracted. The enveloping signal was processed by logarithmic compression and the dynamic range of the image was adjustable. The data were truncated to leave half the depth. The coordinate transformation of the data was carried out and the reference point is the center of the detection window. Then, a horizontal slice was acquired by using morphological processing including an inflation algorithm and a bicubic filling algorithm as shown in Figure 4b.

2.2.3. 3D Imaging

Reconstruction of the 3D breast imaging could be realized by merging the horizontal slices as shown in Figure 4c. The Medical Imaging ToolKit (developed by Institute of Automation of Academia Sinica) was used to realize the 3D visualization.

3. Principles

3.1. Focusing Delay Calculation

The focusing delay time is related to the parameters of the transducer and the position of the focal point. The main parameters involved in the ultrasound transducer include subarray number and array spacing. The subarray number of elements m was 64 and the array spacing d was 1 mm. The position of focal point P was on the central axis of the subarray and the depth of the scanning was 180 mm. The known parameters were the ultrasound velocity c and the sampling rate f_s, which were 1520 m/s and 40 MHz, respectively. The calculation of the focusing delay time τ of the 64 channel linear array transducer is shown in Figure 5. However, the calculation only needed to consider channels 1 to 32, because the 64 channels were symmetrically related to the central axis of the subarray.

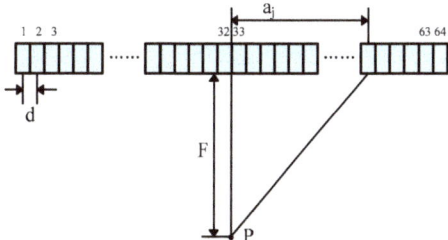

Figure 5. Calculation of the focusing delay time τ for the linear array transducer.

The focusing delay time $\tau_j(F, \beta_j)$ of channel j can be calculated with the formula shown in Equation (1).

$$
\begin{aligned}
a_j &= \left| \tfrac{64+1}{2} - j \right| \times d \quad (j = 1, 2, \ldots, 32) \\
\tau_j(F, \beta_j) &= \left(\sqrt{a_j^2 + F^2} - F \right)/c \\
\Delta F &= c/2f_s = 0.019 \text{mm}
\end{aligned}
\tag{1}
$$

where the distance between array element j and the center axis is a_j; the angle between the array element j and the center axis is β_j; the focal length is F; the ultrasound velocity is c; the array spacing is d; and the minimum distance between the two adjacent focal points is ΔF [31–33].

3.2. Delay Data Analysis

The range of the scanning depth was 2 to 180 mm and the focusing delay data of 32 channels were calculated according to Equation (1). The delay focusing data of channels 1–32 update in real time with the change of the focal depth. The variation of focusing delay data relative to the focal depth F and channel j is shown in Figure 6. For the focal depth F and the channel j, the variation is a monotonically decreasing function. The smaller the distance between two adjacent focal points, the more focusing points means that the quality of the image is higher. Whereas, field programmable gate array (FPGA) consumption also increases. The tradeoff between image quality and FPGA consumption, the distance between two adjacent focal points, was 0.45 mm. These features showed that the following focusing delay data could be stored to some extent that used less memory.

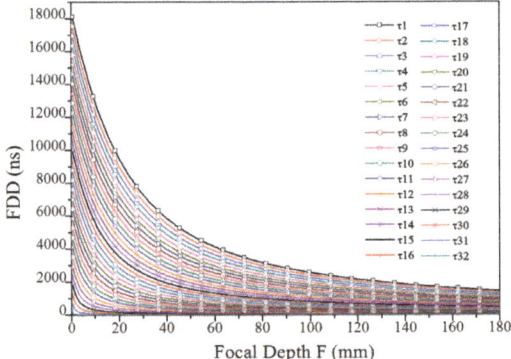

Figure 6. Focusing delay data $\tau_j(F, \beta_j)$ of each channel.

The transmission and receiving of ultrasonic transducers were controlled by the 64 channel ultrasonic signal circuit and the beam synthesis algorithm was realized according to the focusing delay data. By adopting the sequential scanning method, the focus changed dynamically during the scanning process, so that the beam of the whole detection depth could converge well.

4. Results

4.1. Breast Model Imaging

In this section, multi-perspective ultrasound imaging was realized with the cylindrical motion of the linear arrays and the experimental setup was built as shown in Figure 7. The system description is described in Section 2.1. The breast model is made of similar to the mammary gland material, close to the hardness of soft tissue and can be used for ultrasound imaging research. The model with a 2 cm mass in the center and six 1-cm superficial masses were used for experimental analysis as shown in Figure 8d. The breast model size was 15.5 cm × 8 cm. The test was carried out in the water tank and the working temperature was constant at 32 °C. The sound velocity was 1520 m/s and the maximum detection depth was 17.8 cm. The PMUT's center frequency was 3.5 MHz and the element spacing center to center was 1 mm. The 64 channel ultrasonic signal acquisition circuit was used to realize the point-by-point dynamic focusing, where the sampling frequency was 40 MHz. The breast model was placed through the detection window. The rotation interval was set to 2 degrees. When the rotation was 90 degrees, the cylindrical scanning was completed. Hence, 180 × 256 A-scan lines were obtained. These scan lines were analyzed from multiple perspectives according to the algorithm described in Section 2.2 and the results are shown in Figures 8–10. The experimental setup (Figure 7a) can be encapsulated into the detection bed as shown in Figure 7d, which will be used for clinical imaging research in the future.

A total of 180 vertical slices were obtained where the dynamic range of the image was 50 dB and slices with different angles are shown in Figure 10. We could distinguish the boundary of the breast model. The size of the hypothetical masses could be easily seen, which were about 1.9 cm and 0.9 cm, respectively. Compared with the theoretical value, the relative error of the mass size was 5%. Furthermore, the location of the hypothetical masses could be determined by the distance from the center of the mass to the chest wall and the central line of the breast. However, the number of the masses could be determined with some difficulty. As the detection depth increased, the breast boundary became somewhat blurred. This was due to the attenuation problem of the ultrasound in soft tissue transmission. In general, the attenuation coefficient of the soft tissue is 0.6~0.7 dB/(cm/MHz) [12]. As the detection depth increases, the ultrasonic pulse echo signal will be weaker because of the attenuation. However, if the signal attenuation is too fast, the detection depth will be limited.

The acquisition of vertical slices requires a depth detection of 18 cm, which have a uniform resolution in the direction of the depth.

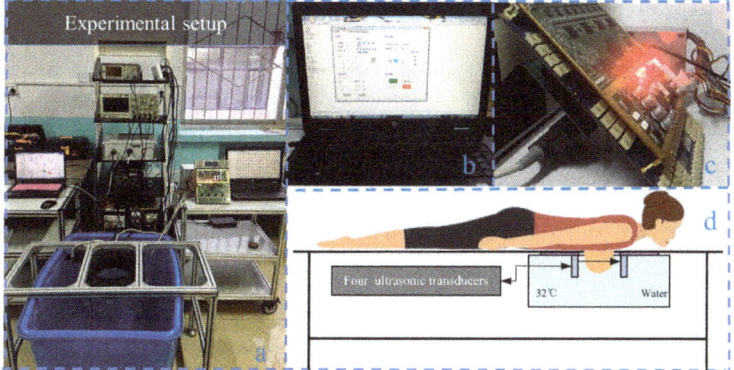

Figure 7. (**a**) Experimental setup. (**b**) PC workstation. (**c**) 64 channel ultrasonic signal receiving and transmitting circuits. (**d**) Diagram of the clinical imaging.

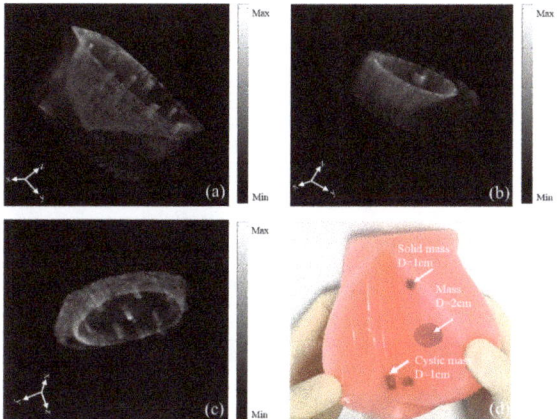

Figure 8. Ultrasonic tomography using this setup. (**a**) The whole breast model imaging. (**b**) The upper part of the breast imaging. (**c**) The lower part of the breast imaging. (**d**) Breast model.

A total of 256 horizontal slices were acquired at different nth elements and representative images are shown in Figure 9. The breast boundary could be seen clearly from the chest wall to the nipple. The masses number could be easily seen. Six small masses could be seen in the superficial surface and one big mass can be seen in the center. The size and location could also be detected. At the same time, the contour of the breast was more complete than the vertical slices, which was due to the horizontal ultrasound tomography algorithm. Hence, this method could reduce the detection depth by half, only requiring a depth detection of 9 cm. Furthermore, the image center resolution was higher than the edge, particularly at the center of the breast where each pixel was viewed from 180 degree angles.

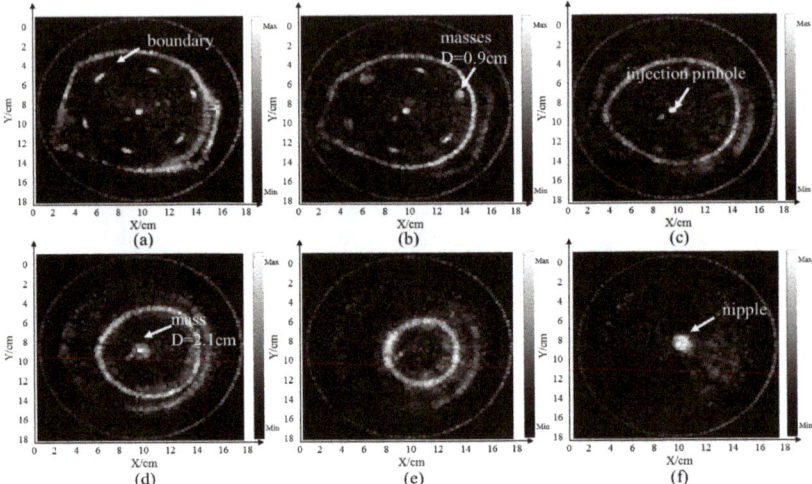

Figure 9. Ultrasonic horizonal tomography images using this setup. (**a**) Slice N = 2. (**b**) Slice N = 15. (**c**) Slice N = 25. (**d**) Slice N = 40. (**e**) Slice N = 50. (**f**) Slice N = 72.

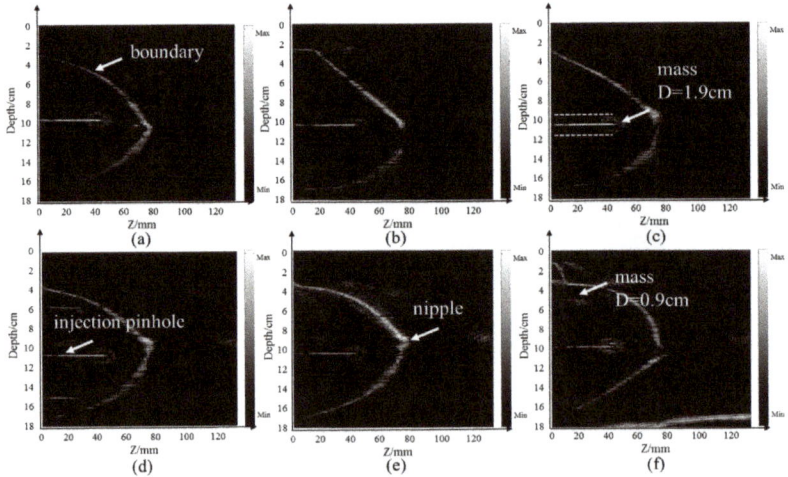

Figure 10. Ultrasonic vertical tomography images using this setup. (**a**) Angle = 24°. (**b**) Angle = 64°. (**c**) Angle = 96°. (**d**) Angle = 164°. (**e**) Angle = 196°. (**f**) Angle = 258°.

The 256 horizontal slices and the 3D ultrasonic imaging of the breast model were realized using the MITK software platform, as shown in Figure 8. The experimental results showed that 3D imaging was more intuitive for tumor detection when compared with the 2D imaging sequences.

5. Discussions and Conclusions

In this project, the breast model was rapidly collected from a series of angles to obtain data from different perspectives. The vertical slices and horizontal slices could be reconstructed retrospectively and the 3D image could be further processed and displayed. The detection of the size, position and shape of different masses was realized with the minimum size of 1 cm. Even the image of the injection pinhole could be seen. Furthermore, each perspective image had its own characteristics. For the vertical slice, the image had a uniform resolution and was relatively easy to obtain. However, it required a larger detection depth. To some extent, the transmitter frequency of the transducer was limited. As for the horizontal slice, the algorithm was relatively complex, with the internal resolution higher than the edge, which is more suitable for the detection of breast deep mass. Meanwhile, it can reduce the demand for transducer detection depth. Thus, it is beneficial to the application of high frequency probes. The 2D slice sequence diagnosis requires more experience for the operator, whereas the information of the breast model and the masses can be shown intuitively through 3D imaging. Therefore, masses can be comprehensively detected from different perspectives and this approach will help to improve the specificity and sensitivity of ultrasonic diagnosis. However, it is still difficult to differentiate between benign over growths in breast tissue (or even calcifications) from malignant tumors. The mechanical and elastic changes in cancerous tissues result in higher density and sound velocity in breast cancer. Mean values of the sound velocity are as follows: fat, 1478 m/s; glandular breast, 1510 m/s; benign breast tumors, 1513 m/s; and malignant breast tumors, 1548 m/s [30]. These data manifest that breast density can be assessed by sound velocity and attenuation. Thus, the multi-perspective imaging technology convergence sound velocity and attenuation imaging algorithm will help with the more specific detection of breast lesions. Meanwhile, the improvement in the sensitivity depends on the study of high density integrated ultrasonic transducer arrays. These methods will overcome the roadblocks in using this approach in the clinic along with biopsies for the breast cancer diagnosis, which will be helpful to reduce the retest rate and improve its accuracy.

Funding: This work was supported by the National Key Research and Development Project under Grants 2016YFC0101900 and 2016YFC0105004 and was sponsored by the Fund for Shanxi '1331 Project' Key Subject Construction.

Acknowledgments: Methodology, C.L.; Software, Y.C.; Formal Analysis, C.L.; Investigation, G.Z.; Data Curation, C.L.; Writing-Original Draft Preparation, C.L.; Writing-Review & Editing, W.Z.; Supervision, C.X. and B.Z.; Funding Acquisition, B.Z., C.Z. and G.Z.

Conflicts of Interest: The authors declare no conflict of interest.

References

1. Grosenick, D.; Rinneberg, H.; Cubeddu, R.; Taroni, P. Review of optical breast imaging and spectroscopy. *J. Biomed. Opt.* **2016**, *21*, 91311. [CrossRef] [PubMed]
2. Stafford, R.J.; Whitman, G.J. Ultrasound Physics and Technology in Breast Imaging. *Ultrasound Clin.* **2011**, *6*, 299–312. [CrossRef]
3. Kamaya, A.; Machtaler, S.; Sanjani, S.S.; Nikoozadeh, A.; Sommer, F.G.; Khuri-Yakub, B.T.; Willmann, J.K.; Desser, T.S. New technologies in clinical ultrasound. *Semin. Roentgenol.* **2013**, *48*, 214. [CrossRef] [PubMed]
4. Duric, N.; Littrup, P.; Li, C.; Roy, O.; Schmidt, S.; Cheng, X.; Seamans, J.; Wallen, A.; Bey-Knight, L. Breast imaging with SoftVue: Initial clinical evaluation. *Proc. Spie Int. Soc. Opt. Eng.* **2014**, *9040*, 382–385.
5. Sasieni, P.D.; Shelton, J.; Ormiston-Smith, N.; Thomson, C.S.; Silcocks, P.B. What is the lifetime risk of developing cancer?: The effect of adjusting for multiple primaries. *Br. J. Cancer* **2011**, *105*, 460–465. [CrossRef] [PubMed]
6. Pisano, E. Diagnostic performance of digital versus film mammography for breast-cancer screening. *N. Eng. J. Med.* **2005**, *353*, 1773–1783. [CrossRef] [PubMed]
7. Hylton, N. Magnetic Resonance Imaging of the Breast: Opportunities to Improve Breast Cancer Management. *J. Clin. Oncol.* **2005**, *23*, 1678–1684. [CrossRef] [PubMed]
8. Lord, S.J.; Lei, W.; Craft, P.; Cawson, J.N.; Morris, I.; Walleser, S.; Griffiths, A.; Parker, S.; Houssami, N. A systematic review of the effectiveness of magnetic resonance imaging (MRI) as an addition to mammography and ultrasound in screening young women at high risk of breast cancer. *Eur. J. Cancer* **2007**, *43*, 1905–1917. [CrossRef] [PubMed]
9. Christiansen, C.L.; Wang, F. Predicting the Cumulative Risk of False-Positive Mammograms. *J. Natl. Cancer Inst.* **2000**, *92*, 1657–1666. [CrossRef]
10. Holländer, B.; Hendriks, G.A.; Mann, R.M.; Hansen, H.H.; de Korte, C.L. Plane-Wave Compounding in Automated Breast Volume Scanning: A Phantom-Based Study. *Ultrasound Med. Biol.* **2016**, *42*, 2493–2503. [CrossRef]

11. Ruiter, N.V. GPU based 3D SAFT reconstruction including phase aberration. In Proceedings of the Medical Imaging 2014: Ultrasonic Imaging and Tomography, International Society for Optics and Photonics, San Diego, CA, USA, 15–20 February 2014; Volume 10, pp. 252–260.
12. Wiskin, J.; Borup, D.T.; Johnson, S.A.; Berggren, M. Non-linear inverse scattering: High resolution quantitative breast tissue tomography. *J. Acoust. Soc. Am.* **2012**, *131*, 3802. [CrossRef] [PubMed]
13. Ruiter, N.V.; Zapf, M.; Hopp, T.; Dapp, R.; Kretzek, E.; Birk, M.; Kohout, B.; Gemmeke, H. 3D ultrasound computer tomography of the breast: A new era? *Eur. J. Radiol.* **2012**, *1*, 133–134. [CrossRef]
14. Terada, T.; Yamanaka, K.; Suzuki, A.; Tsubota, Y.; Wu, W.; Kawabata, K.I. Highly precise acoustic calibration method of ring-shaped ultrasound transducer array for plane-wave-based ultrasound tomography. *Jpn. J. Appl. Phys.* **2017**, *56*, 07JF07. [CrossRef]
15. Oeri, M.; Bost, W.; Tretbar, S.; Fournelle, M. Calibrated Linear Array-Driven Photoacoustic/Ultrasound Tomography. *Ultrasound Med. Biol.* **2016**, *42*, 2697–2707. [CrossRef] [PubMed]
16. Koch, A.; Stiller, F.; Lerch, R.; Ermert, H. An ultrasound tomography system with polyvinyl alcohol (PVA) moldings for coupling: In vivo results for 3-D pulse-echo imaging of the female breast. *Ultrason. Ferroelectr. Freq. Control IEEE Trans.* **2015**, *62*, 266–279. [CrossRef]
17. Zhang, X.; Fincke, J.; Kuzmin, A.; Lempitsky, V.; Anthony, B. A single element 3D ultrasound tomography system. In Proceedings of the IEEE Engineering in Medicine and Biology Society, Milan, Italy, 25–29 August 2015; pp. 5541–5544.
18. Tasinkevych, J.; Trots, I. Circular Radon Transform Inversion Technique in Synthetic Aperture Ultrasound Imaging: An Ultrasound Phantom Evaluation. *Arch. Acoust.* **2014**, *39*, 569–582. [CrossRef]
19. Nebeker, J.; Nelson, T.R. Imaging of sound speed using reflection ultrasound tomography. *J Ultrasound Med* **2012**, *31*, 1389–1404. [CrossRef]
20. Sandhu, G.Y.; Li, C.; Roy, O.; Schmidt, S.; Duric, N. Frequency Domain Ultrasound Waveform Tomography: Breast Imaging Using a Ring Transducer. *Phys. Med. Biol.* **2016**, *60*, 5381–5398. [CrossRef]
21. Sanpanich, A.; Hamamoto, K.; Pintavirooj, C. An investigation on attenuation UCT with wave paths enhancement for breast ultrasound. *Ieej Trans. Electrical Electronic Eng.* **2012**, *7*, S105–S113. [CrossRef]
22. Rouyer, J.; Mensah, S.; Franceschini, E.; Lasaygues, P.; Lefebvre, J.P. Conformal ultrasound imaging system for anatomical breast inspection. *IEEE. Trans. Ultrason. Ferroelectr. Freq. Control* **2012**, *59*, 1457–1469. [CrossRef]
23. Birk, M.; Dapp, R.; Ruiter, N.V.; Becker, J. GPU-based iterative transmission reconstruction in 3D ultrasound computer tomography. *J. Parallel Distrib. Comput.* **2014**, *74*, 1730–1743. [CrossRef]
24. Jaeger, M.; Held, G.; Peeters, S.; Preisser, S.; Grünig, M.; Frenz, M. Computed ultrasound tomography in echo mode for imaging speed of sound using pulse-echo sonography: Proof of principle. *Ultrasound Med. Biol.* **2015**, *41*, 235–250. [CrossRef] [PubMed]
25. Sandhu, G.Y.; West, E.; Li, C.; Roy, O.; Duric, N. 3D frequency-domain ultrasound waveform tomography breast imaging. In Proceedings of the Medical Imaging 2017: Ultrasonic Imaging and Tomography. International Society for Optics and Photonics, Orlando, FL, USA, 11–16 February 2017; p. 1013909.
26. Yu, S.; Wu, S.; Zhuang, L.; Wei, X.; Sak, M.; Neb, D.; Hu, J.; Xie, Y. Efficient Segmentation of a Breast in B-Mode Ultrasound Tomography Using Three-Dimensional GrabCut (GC3D). *Sensors* **2017**, *17*, 1827. [CrossRef] [PubMed]
27. Kim, Y.C.; Choi, J.Y.; Chung, Y.E.; Bang, S.; Kim, M.J.; Park, M.S.; Kim, K.W. Comparison of MRI and endoscopic ultrasound in the characterization of pancreatic cystic lesions. *Ajr Am. J. Roentgenol* **2010**, *195*, 947–952. [CrossRef] [PubMed]
28. Torres, F.; Fanti, Z.; Cosío, F.A. 3D freehand ultrasound for medical assistance in diagnosis and treatment of breast cancer: Preliminary results. In Proceedings of the IX International Seminar on Medical Information Processing & Analysis, Mexico City, Mexico, 11–14 November 2013.
29. Kretzek, E.; Zapf, M.; Birk, M.; Gemmeke, H.; Ruiter, N.V. GPU based acceleration of 3D USCT image reconstruction with efficient integration into MATLAB. In Proceedings of the Medical Imaging: Ultrasonic Imaging, Tomography, & Therapy, Lake Buena Vista (Orlando Area), FL, USA, 9–14 February 2013.
30. Liu, C.; Xue, C.; Zhang, B.; Zhang, G.; He, C. The Application of an Ultrasound Tomography Algorithm in a Novel Ring 3D Ultrasound Imaging System. *Sensors* **2018**, *18*, 1332. [CrossRef] [PubMed]
31. Yin, J.; Tao, C.; Liu, X. Dynamic focusing of acoustic wave utilizing a randomly scattering lens and a single fixed transducer. *J. Appl. Phys.* **2017**, *121*, 1606. [CrossRef]

32. Wang, P.; Jiang, J.; Luo, H.; Li, F.; Sun, G.; Cui, S. The research of compression and generation of high-precision dynamic focusing delay data for ultrasound beamformer. *Clust. Comput.* **2017**, *20*, 1–11. [CrossRef]
33. U-Wai, L.; Gang-Wei, F.; Pai-Chi, L. Lossless data compression for improving the performance of a GPU-based beamformer. *Ultrason. Imaging* **2015**, *37*, 135–151.

© 2019 by the authors. Licensee MDPI, Basel, Switzerland. This article is an open access article distributed under the terms and conditions of the Creative Commons Attribution (CC BY) license (http://creativecommons.org/licenses/by/4.0/).

Article

Speckle Reduction on Ultrasound Liver Images Based on a Sparse Representation over a Learned Dictionary

Mohamed Yaseen Jabarulla and Heung-No Lee *

School of Electrical Engineering and Computer Science, Gwangju Institute of Science and Technology, Gwangju 61005, Korea; yaseen@gist.ac.kr
* Correspondence: heungno@gist.ac.kr

Received: 20 April 2018; Accepted: 28 May 2018; Published: 31 May 2018

Abstract: Ultrasound images are corrupted with multiplicative noise known as speckle, which reduces the effectiveness of image processing and hampers interpretation. This paper proposes a multiplicative speckle suppression technique for ultrasound liver images, based on a new signal reconstruction model known as sparse representation (SR) over dictionary learning. In the proposed technique, the non-uniform multiplicative signal is first converted into additive noise using an enhanced homomorphic filter. This is followed by pixel-based total variation (TV) regularization and patch-based SR over a dictionary trained using K-singular value decomposition (KSVD). Finally, the split Bregman algorithm is used to solve the optimization problem and estimate the de-speckled image. The simulations performed on both synthetic and clinical ultrasound images for speckle reduction, the proposed technique achieved peak signal-to-noise ratios of 35.537 dB for the dictionary trained on noisy image patches and 35.033 dB for the dictionary trained using a set of reference ultrasound image patches. Further, the evaluation results show that the proposed method performs better than other state-of-the-art denoising algorithms in terms of both peak signal-to-noise ratio and subjective visual quality assessment.

Keywords: ultrasound; speckle reduction; medical image processing; sparse representation; K-singular value decomposition; dictionary learning; B-mode imaging

1. Introduction

In the last 20 years, there has been growing interest in the use of ultrasound imaging for a variety of applications, such as observing the blood flow through an organ or other structures; determining bone density; imaging the heart, a fetus, or ocular structures; or diagnosing cancers [1,2]. Ultrasound imaging has been widely applied owing to its ability to produce real-time images and videos. Ultrasound images are captured in real-time by transmitting high frequency sound waves through body tissue. It comprises an array of transducer elements that sequentially echo the signal for each spatial direction to generate a raw line signal. The scan is converted to construct a Cartesian image from the processed raw line signal [2].

In recent years, many researchers have attempted to develop computer-aided diagnostic (CAD) systems for diagnosing liver and breast cancers [3–6] based on ultrasound imaging. The aim of these systems is to differentiate benign and malignant lesion tissues as well as cysts [7]. A CAD system carries out the diagnosis in four stages: data preprocessing, image segmentation, feature extraction, and classification [4]. Data preprocessing is the first and most vital step in the CAD system process because it reconstructs an image without eliminating the important features by reducing signal-dependent multiplicative noise called speckle [8].

The development of a precise speckle reduction model is an important step to achieve efficient denoising filter design. Recent review articles [4,9], reported that speckle reduction filters are categorized into two broad approaches: spatial filtering and multiscale methods. Techniques under spatial domain filtering include enhanced Frost filtering [10], Lee filtering [11], mean filtering [12], Wiener filtering [13], Kuan filtering [14], and median filtering [15]. Spatial filters utilize local statistical properties to reduce speckle noise. However, small details may not be preserved [9]. Several methods [16–19] use multiscale filtering, which uses the wavelet transform to preserve the image signal regardless of its frequency content. Donoho et al. [20] proposed reducing noise in the wavelet domain by soft thresholding. However, their approach lacked translation invariance when using the discrete wavelet transform. This is resolved by eliminating up and down samplers in the wavelet transform by using a stationary wavelet transform [21], which is a redundant technique because the number of input and output samples at each level is the same. A multiresolution technique called translation invariant image enhancement was proposed in [22]. The proposed technique incorporates noise reduction and directional filtering. Directional filtering is executed using eigenvalues by analyzing the structure of each pixel's neighborhood. Rudin et al. [23,24] and Perona et al. [25] proposed successful image denoising techniques called total variation (TV) and anisotropic smoothing, respectively. These models were improved and extended upon in later works [26,27]. However, all these methods are computationally expensive. In recent years, more efficient denoising techniques such as sparse representation (SR) have been proposed [28–31]. In digital image processing, many signals are sparse; i.e., they contain many coefficients either equal to or close to zero in a specific domain. The objective of SR is to efficiently reconstruct the signal with a linear combination of a few dictionary atoms from the transformed signal domain [32].

This study was conducted with the objective of developing filtering algorithm that can reduce noise without losing significant features or eliminating edges. To this end, this paper proposes, a technique that reduces the speckle noise in ultrasound imaging systems by applying a relatively new signal reconstruction model known as SR [32] to deal with complicated noise properties. Sparse representation provides superior estimation even in an ill-conditioned system [33], and has been found to be very useful in medical imaging applications. However, one challenge of designing this system is the presence of a multiplicative speckle signal because dictionary learning methods are not effective on multiplicative and correlated noise. We overcome this by using two different methods. Firstly, the speckle noise is transformed into additive noise using an enhanced homomorphic filter that can also capture high and low frequency signal of the image. Secondly, we introduced TV regularization of the image and sparse prior over learned dictionaries. Total variation regularization is efficient for noisy image, while the patch-based dictionaries are well adapted to texture features [34], and reduces the artifacts in smooth pixel regions [35]. The advantage of the sparse prior is that it utilizes fewer dictionary columns to reconstruct a noiseless ultrasound image without losing many important features of the signal. Therefore, in our proposed model we combined the two approaches, the patch-based SR over learned dictionaries and the pixel-based TV regularization method, for efficient speckle reduction. The K-singular value decomposition (KSVD) algorithm [36] is used to learn two modified dictionaries from reference ultrasound image datasets and the corrupted images; these are referred to as dictionaries 1 and 2, respectively. The results are evaluated on both dictionaries and compared with conventional algorithms to show that the speckle noise is suppressed effectively in the ultrasound image using SR.

The rest of the paper is organized as follows. Noise model and related works are described in Section 2. The proposed SR framework for speckle reduction in ultrasound imaging is presented in Section 3. In Section 4, the experiments and results obtained are discussed. The paper is concluded in Section 5.

2. Background

2.1. Ultrasound Noise Model

Ultrasound imaging system are often affected by multiplicative speckle [37]. Scattering time differences lead to constructive and destructive interference of the ultrasound pulses that are reflected from biological tissues. Speckle patterns can be classified depending on the spatial distribution, number of scatters per resolution cell, and properties of the imaging system [9]. Speckle noise affects the detectability of the target and reduces the contrast and resolution of the images, making it difficult for a clinician to provide a diagnosis.

In ultrasound, the multiplicative noise models are based on the product of the original signal and noise. Thus, the intensity of a noisy signal depends on the original image intensity. The mathematical expression for a multiplicative speckle model is given by

$$y(i,j) = x(i,j)h(i,j), \qquad (1)$$

where $y(i,j)$ is the speckled image, $x(i,j)$ is the original image, and $h(i,j)$ is the speckle noise. The spatial location of an image is represented using indexes i and j, where index i ranges from 1 to N, and index j from 1 to M.

2.2. Related Work on Multiplicative Noise Reduction

Several algorithms have been proposed to deal with more complex multiplicative and additive speckle noise models [38]. For instance, the Kuan, Frost, Lee filters, and speckle reducing anisotropic diffusion (SRAD) filter [39] are effective on the multiplicative noise model. Other filters, specifically the median, Wiener, and wavelet filters [40], are designed for the additive noise model [4]. However, each filter has certain advantages and limitations [38]. In a few filter models, the quality of the processed image is affected by the window size: large window sizes cause image blurring, degrading the fine details of an image. Conversely, small window sizes do not denoise the image sufficiently. Other widely used multiplicative noise reduction algorithms are based on the TV regularization term [23,41], nonlocal methods [42,43], and wavelet-based approaches [16]. Total variation-based methods effectively remove flat-region-based noise and preserve the edges of images. However, fine details are lost because of over-smoothed textures. Nonlocal algorithms depend on similarities of image patches. Their performance is limited by dissimilar image patches. However, wavelet-based approaches preserve texture information better than TV-based methods. This approach assumes that images in the SRs are based on a fixed dictionary [29,36]. However, certain characteristics of the processed image might not be captured because the dictionary does not contain any similar image content.

To overcome the above disadvantages, over the past few years, researchers have sought to develop an algorithm based on SR in the field of image and signal processing [32]. This is because the pattern similarities of image signals such as textures and flat regions, mean that the signal can be efficiently approximated as a linear combination using a dictionary of only a few functions called atoms [29,34,36]. Elad and Aharon [36] proposed an image denoising algorithm using an adaptive dictionary called KSVD that is based on sparse and redundant representations. It includes sparse coding and dictionary atoms that are updated to better fit the data. The advantage of KSVD compared to fixed dictionaries is that it is effective at removing additive Gaussian noise using the linear combinations of a few atoms, by learning a dictionary from noisy image patches and then reconstructing each patch.

A dictionary $A \in \mathbb{R}^{N_r \times N_c}$, composed of N_c columns of N_r elements, is called a sparse-land model [36]. K-singular value decomposition seeks the best signal representation of image signal y from the sparsest representation α:

$$\widehat{\alpha} = \mathrm{argmin} \|\alpha\|_0 \text{ subject to } \|y - A\alpha\|_2 \leq \varepsilon,$$

where the vectorization of $y(i,j)$ is denoted by vector $y \in \mathbb{R}^{M \times 1}$ and ε is the few number of non-zero entries in α. K-singular value decomposition replaces the dictionary update and sparse coding stages with a simple singular value decomposition. The orthogonal matching pursuit (OMP) method [44] is an effective method to find the sparse approximation. In the OMP, if the noise level is below the approximation, the image patches are rejected. The singular value decomposition constructs better atoms by combining patches to reduce noise for ultrasound speckle reduction. K-singular value decomposition has also proved to effectively reduce the speckle produced by additive white Gaussian noise on corrupted images [29,36].

The filtering algorithm comprises two steps. First, the dictionary is trained from a set of image data patches or from noisy image patches based on KSVD. The next step uses $\widehat{\alpha}$ to compute SR using dictionary A and denoises the image [29].

The method proposed in [45] also uses a dictionary learning approach for denoising ultrasound images. A homomorphic filter is used to convert multiplicative noise into additive white Gaussian noise and then the noiseless signal is reconstructed over image patches (atoms) to create the SR from a learned dictionary. However, noise in flat regions still exists and poor edges make the reconstructed images difficult to analyze. In [34], the authors proposed an image denoising technique that operates directly on multiplicative noise and is based on three terms: SR over an adaptive dictionary, a TV regularization term, and a data-fidelity term. However, the proposed model is nonconvex because of the product between the unknown dictionary and sparse coefficients and the data-fidelity term is a log function. Therefore, solving the squared l_2 norm is difficult. This optimization problem is overcome by the split Bregman technique. However, these methods do not contain high- and low-frequency components of the image. We obtain this information using an enhanced homomorphic filter designed to improve the final image. Furthermore, we utilize the advantages of combining a TV regularization term and SR learned over two modified dictionaries.

3. Sparse Representation Framework for Speckle Reduction

As discussed above, we define our proposed scheme for ultrasound speckle reduction by considering the multiplicative noise model [37] obtained by an ultrasound transducer. Equation (1) can thus rewritten as

$$y_\partial(i,j) = x_\Re(i,j) n_\sigma(i,j), \tag{2}$$

where $y_\partial(i,j)$ is the degraded B-mode image signal [46], $x_\Re(i,j)$ represents the ideal image that must be recovered, and $n_\sigma(i,j)$ represents the speckle noise, generally modelled as a Rayleigh probability density function with random variables [11,47]. Each term includes coordinates (i,j) defined according to the acquisition geometry.

In general, a homomorphic filter [48] is a well-proven technique for converting multiplicative noise. In this study, we modified it by taking the log of the multiplicative noisy signal and filtering the image using a Butterworth high-pass (BW-HP) filter to attenuate low frequencies in the transmitted signal while preserving the high frequencies in the reflected component. The equation of the BW-HP filter is

$$H_B(u,v) = \frac{1}{1 + \left[D_0/\sqrt{u^2+v^2}\right]^{2f}}, \tag{3}$$

where, D_0 is the cut-off frequency and f is the order of the filter. We varied the frequency values u and v of the i and j spatial coordinates. We used the BW-HP filter because it generates fewer ringing artifacts on the image signal.

We also used a Gaussian low pass (GLP) filter to smooth the low-frequency signal component in the log domain. The equation of the GLP filter is

$$H_G(u,v) = e^{-D^2(u,v)/2D_0^2}, \tag{4}$$

where $D(u,v)$ is the distance from the origin in the frequency plane. Finally, the additive noise signals were estimated by applying inverse transform.

Figure 1 shows the steps used to convert an original noisy image into an image with additive noise using the enhanced homomorphic transform. This technique consists of five steps. We first take the log on both sides of Equation (2) and use a two-dimensional fast Fourier transform (FFT) to represent the image in the frequency domain. Then, the Fourier image is filtered with two filter functions, those are the BW-HP and GLP filters [12]. The BW-HP filter increases the contrast of the image signal corresponding to the high-frequency component. The GLP filter smooths the noise signal without eliminating the entire low-frequency component. Both filtered signals are applied to the two-dimensional inverse fast Fourier transform (IFFT). Finally, taking the exponent of the image, we obtain the transformed image. This process is discussed in detail below.

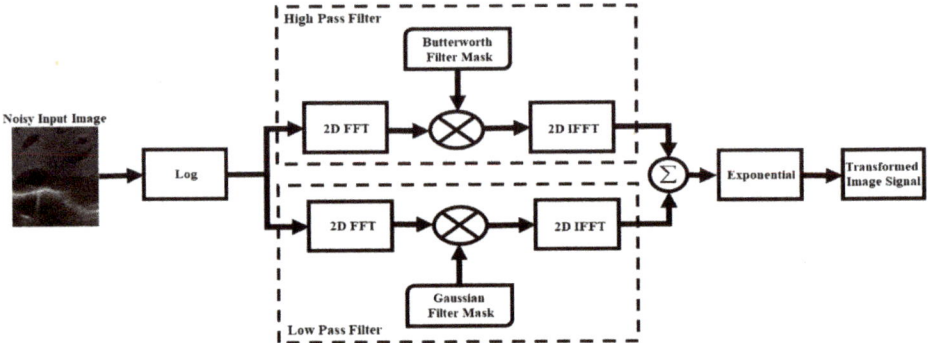

Figure 1. Flow diagram of the enhanced homomorphic filter. FFT: fast Fourier transform; IFFT inverse fast Fourier transform.

Step 1: Take the log on both sides of the $x_\Re(i,j)$ and the $n_\sigma(i,j)$ signal; now the multiplicative noise can written as

$$\log(y_\partial(i,j)) = \log(x_\Re(i,j)) + \log(n_\sigma(i,j)), \qquad (5)$$

After being transformed logarithmically, the signal now contains Gaussian additive noise [49]. We remove $\log(x_\Re(i,j))$ from the speckled ultrasound image by applying an additive noise suppression algorithm. Thus, the problem is now to estimate $\log(x_\Re(i,j))$ from noisy data.

Step 2: Apply FFT to convert the image into the frequency domain. Equation (5), thus becomes,

$$y_\partial(u,v) = F_{x_\Re}(u,v) + F_{n_\sigma}(u,v), \qquad (6)$$

where, $F_{x_\Re}(u,v)$ and $F_{n_\sigma}(u,v)$ are the FFT of $\log(x_\Re(i,j))$ and $\log(n_\sigma(i,j))$, respectively.

Step 3: Apply BW-HP and GLP to the $y_\partial(u,v)$ by means of two filter function $H_B(u,v)$ and $H_G(u,v)$ from Equations (3) and (4) respectively in the frequency domain. The filtered version of $S(u,v)$ is written as

$$S(u,v) = H_B(u,v)y_\partial(u,v) + H_G(u,v)y_\partial(u,v). \qquad (7)$$

Step 4: Take the inverse Fourier transform of Equation (7) to get the converted signal in the spatial domain

$$\overline{S}(i,j) = \mathcal{F}^{-1}\{S(u,v)\}.$$

Step 5: Finally, we obtain the transformed image $t(i,j)$ by taking the exponent of the image using the following equation

$$t(i,j) = \exp\{\overline{S}(i,j)\}.$$

In this paper, we model the transformed image as additive noise degradation $W(i,j)$ of the original image $x_\Re(i,j)$, i.e.,

$$t(i,j)x_\Re(i,j) + W(i,j). \qquad (8)$$

This completes how we have used the homomorphic filter to transform the speckle noise into additive noise. The two filter functions are utilized to improve edge information by enhancing contrast and smooths the additive noise of the transformed image.

Figure 2 shows the output of the enhanced homomorphic filter at the BW-HP and GLP filter stages. It is clear that the image in Figure 2b has an increased intensity because the low frequency signal is attenuated and the image in Figure 2c is smoothed by the GLP filter. The sum of these two signals is the final transformed noisy image.

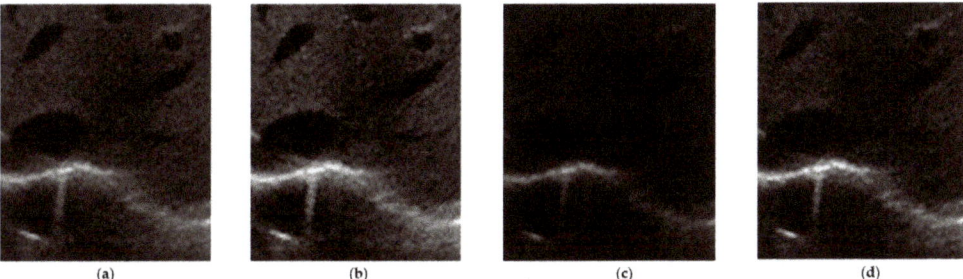

Figure 2. (**a**) Noisy ultrasound image; (**b**) Butterworth high-pass (BW-HP) filtered image; (**c**) Gaussian low pass (GLP) filtered image; and (**d**) transformed output of ultrasound noisy image.

An ultrasound image $x_\Re(i,j)$ can be represented as sparse in the gradient domain. We thus define here a difference signal. A pixel-based TV regularization can be performed on the transformed image for more effective denoising. The horizontal and vertical difference matrices are defined below [50].

$$V_i x_\Re(i,j) = \begin{cases} x_\Re(i+1,j) - x_\Re(i,j), & \text{if } i < n \\ 0, & \text{if } i = n \end{cases}$$

$$V_j x_\Re(i,j) = \begin{cases} x_\Re(i,j+1) - x_\Re(i,j), & \text{if } j < m \\ 0, & \text{if } j = m \end{cases}$$

Further, the difference signal of $x_\Re(i,j)$ is defined as

$$V_{i,j} x_\Re(i,j) = \begin{pmatrix} V_i x_\Re(i,j) \\ V_j x_\Re(i,j) \end{pmatrix}.$$

We can show that there exists a dictionary $A \in \mathbb{R}^{N_r \times N_c}$ with which the original image can be sparsely represented as

$$x_\Re = A\alpha,$$

where $x_\Re$ is the vectorization of the recovered signal $x_\Re(i,j)$ such that $x_\Re \in \mathbb{R}^{N_r}$. If a signal $x_\Re$ is K-sparse in the dictionary $A \in \mathbb{R}^{N_r \times N_c}$ for $N_c > N_r$, we imply that the signal can be represented with K columns of the dictionary. The column vector $\alpha \in \mathbb{R}^{N_c \times 1}$ is the vector of the coefficients. Then, by optimizing the following convex problem, the signal $x_\Re$ can be recovered:

$$\begin{aligned} &\min \|\alpha\|_0, \\ &\text{subject to } \|t - A\alpha\|_2^2 \leq \varepsilon. \end{aligned} \qquad (9)$$

In Equation (9), a $NM \times 1$ column vector t is the vectorization of the transformed image $t(i,j)$, note that $NM = N_r$. Also note that ε is a utility parameter selectable according to the noise strength.

This convex constrained problem can be transformed into an unconstrained optimization problem using the Lagrange multiplier method [51]:

$$\min \|t - A\alpha\|_2^2 + \tau \|\alpha\|_0. \tag{10}$$

Using the unconstrained problem, we are able to combine a regularization term, which is weighted by parameter $\tau > 0$ and a quadratic data-fidelity term. Equation (10) is not ready for use yet since we do not know the sparsity dictionary A. Therefore, we use the following approach where the dictionary, the sparse representation coefficient vector α, and the image vector $x_\Re$ are estimated altogether. The overall optimized discrete sparse model proposed in this paper, for denoising the ultrasound image, can be written as

$$\left\{ \hat{x}_\Re, \widehat{\alpha}_{ij}, \widehat{A} \right\} = \min_{x_\Re, \alpha_{ij}, A} \lambda \|V x_R\|_1 + \tau \sum_{ij} \|R_{ij} t - A \alpha_{ij}\|_2^2 + \tau \sum_{ij} \|\alpha_{ij}\|_0, \tag{11}$$

where R_{ij} is an operation that extracts a square image patch from the transformed image t located at the i, j pixels of the image. The notation $\|.\|_1$ is used to imply the l_1 norm, which is the sum of the absolute values of the argument signal, which in this case is the difference signal $V x_R$. There are two positive parameters λ and τ used to balance the contribution of different terms. In Equation (11), the first and second terms are the TV regularization norm and the sparse representation prior. Optimization in Equation (11) seeks to find a solution with which each patch of the recovered image can be represented by a dictionary matrix with sparse coefficient α in the sense of a bounded error. The l_0 norm gives the sparsity constraint which controls the sparsity coefficients of any small image patch.

As mentioned in Related Work Section 2.2, there is a sparse coding stage that utilizes the KSVD iterative process. In the first stage, sparse coding is performed assuming fixed $x_\Re$ and A. In the second stage, dictionary A is updated to minimize using known sparse coefficients α and $x_\Re$. The sparse coefficients $\widehat{\alpha}_{ij}$ are computed using the OMP method [52] because of its efficiency and simplicity. Elad et al. [29] showed that learning a dictionary trained from good quality image patches and noisy images results in better performance.

In this paper, we use two approaches to train the dictionary. The first approach is to use a group of image patches taken from many ultrasound reference images. We call the dictionary obtained from this approach Dictionary 1. The second approach is to use the corrupted images and call them Dictionary 2. We aim to compare the performance difference based on these two approaches. The comparison is made in the Results section.

It should be noted that Equation (11) is non-convex because of the non-differentiable TV regularization term and the product of the unknowns A and α_{ij}. We overcome this by using the split Bregman iterative approach [53].

Overall, the proposed algorithm can be summarized as follows:

1. Convert the multiplicative noise into additive noise using an enhanced homomorphic filter and capture the high- and low-frequency components to retain detailed information.
2. Apply pixel-based TV regularization to smooth the filtered image signal.
3. Apply patch-based sparse representation over a dictionary trained using the KSVD algorithm. We employed two modified dictionaries—one trained with a set of reference ultrasound image patches and another trained using the speckled image patches.
4. Iterate between the TV regularization and sparse representation procedure to improve the reconstructed image.

Figure 3 summarizes the proposed algorithm.

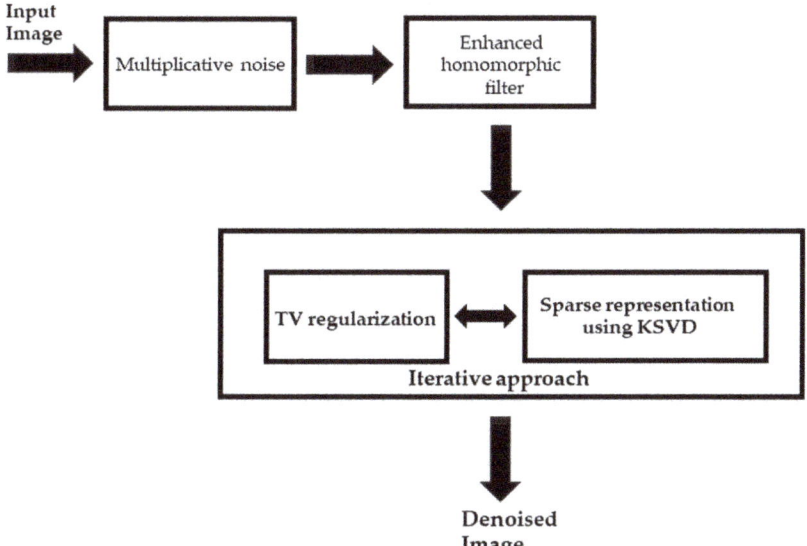

Figure 3. Proposed despeckle model for an ultrasound image. KSVD: K-singular value decomposition.

3.1. Performance Estimation

The reconstructed denoised image using the proposed algorithm were compared with the original image. Two image quality metrics were used for quantitative performance measurements: peak signal-to-noise ratio (PSNR) and mean structural similarity (MSSIM) [54]. Peak signal-to-noise ratio is defined as:

$$PSNR = 10 \log_{10} \frac{N_{max}}{\frac{1}{MN} \sum_{n=1}^{N} \sum_{m=1}^{M} |x(n,m) - \widehat{x}(n,m)|^2}, \quad (12)$$

where N_{max} represents the maximum fluctuations in the input image. Here, $N_{max} = (2^n - 1)$, $N_{max} = 255$, when the components of a pixel are encoded using eight bits. N denotes the number of pixels processed, $x(n,m)$ is the original signal, and $\widehat{x}(n,m)$ is the recovered image signal. In $MSSIM$, the structures of the two images are compared after normalizing the variance and subtracting the luminance as follows:

$$MSSIM = \frac{1}{N} \sum_{i=1}^{N} \left[l(\widehat{x},x) \right]^{\alpha} \cdot \left[c(\widehat{x},x) \right]^{\beta} \cdot \left[s(\widehat{x},x) \right]^{\gamma}, \quad (13)$$

where $l(\widehat{x},x)$ denotes luminance, $c(\widehat{x},x)$ denotes contrast, and $s(\widehat{x},x)$ denotes structure comparison functions. Further, α, β, and r are weighted parameters that are used to adjust the relative importance of the three components.

4. Experimental Results and Discussion

4.1. Simulations on Synthetic Images

In this section, we analyze the performance of the proposed approach on the synthetic Shepp–Logan phantom test image [55] (Figure 4a) with a speckle noise variance of $\sigma = 10$ (Figure 4b) of a 256 × 256 pixel size. This result helps us to understand the effectiveness of the simulated image, clearly determine the distinctive features of the image, and optimize the algorithm before testing on the clinical datasets. We compared the proposed algorithm with some standard speckle reduction

Appl. Sci. 2018, 8, 903

filters for ultrasound liver images [4]. The compared algorithms were local statistical filters such as the Frost filter [10], Lee filter [11], 3 × 3 Weiner filter [13], Kuan filter [14], 3 × 3 median filter [15], and speckle reducing anisotropic diffusion (SRAD) filter [39]. In addition, multiscale filters such as wavelets [40] were evaluated. The despeckled images in Figure 4e–g show that the Frost, wavelet, and Kuan filters do not effectively reduce noise. In contrast, Figure 4h–j show that the median, Weiner, and SRAD filters, reduce most noise; however, the edges are not preserved and artificial noises can be introduced to a certain extent. This result verifies that the proposed SR technique reduces noise and preserves the edges better than the conventional methods on synthetic images. Table 1 shows the PSNR value and MSSIM value. The proposed algorithm reconstructs the original image with a PSNR value of 36.86 dB with Dictionary 1 and 37.04 dB with Dictionary 2.

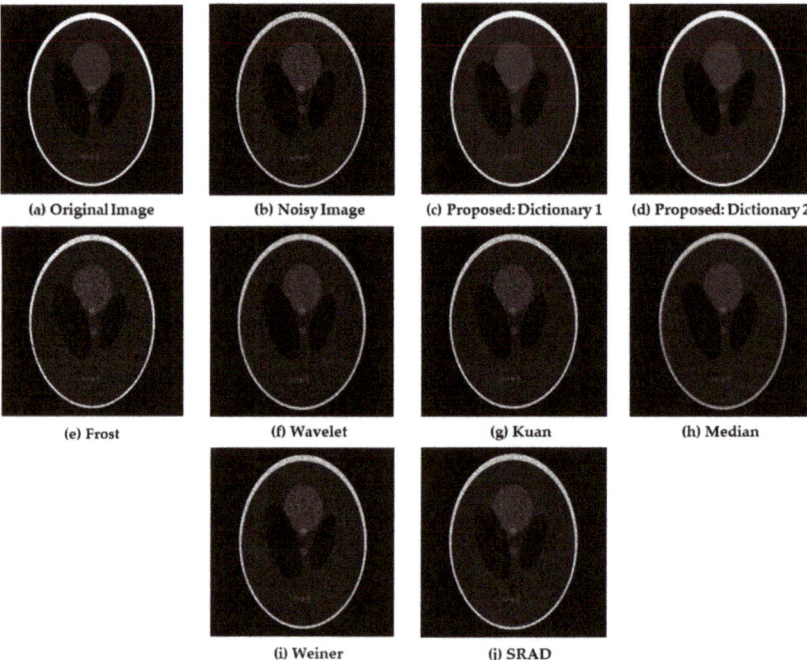

Figure 4. (**a**) Original image; (**b**) noisy image. Results of the proposed method with (**c**) Dictionary 1 and (**d**) Dictionary 2; Results of the (**e**) Frost; (**f**) wavelet; (**g**) Kuan; (**h**) median; (**i**) Weiner; and (**j**) speckle reducing anisotropic diffusion (SRAD) filters.

Table 1. Peak signal-to-noise ratio (PSNR) and mean structural similarity (MSSIM) for the synthetic images for σ = 10.

Models	PSNR (dB)	MSSIM
Noise image	32.113	0.727
Frost	32.466	0.768
Wavelet	33.214	0.801
Kuan	32.895	0.794
Median	34.597	0.839
SRAD	33.434	0.827
Weiner	33.782	0.834
Proposed: Dictionary 1	36.862	0.953
Proposed: Dictionary 2	37.044	0.967

4.2. Clinical Liver Ultrasound Images

The proposed algorithm efficiency was estimated using a set of B-mode greyscale ultrasound liver images. The images were obtained using the ECUBE 12R ultrasound research system from Alpinion medical systems, Seoul, Korea. The components used to generate the ultrasound images include a 128-element linear transducer at a center frequency of 5 MHz, a lateral beam width of 1.5 mm, and a pulse length of 1 mm. In our experiment, sparse coding was performed using two dictionaries with a 64×256 size, designed to handle patches of 8×8 size pixels ($N = 64$ and $K = 256$)—one trained from a noisy image and the other trained from a set of reference images.

The training data were constructed from a dataset comprising 3245 reference ultrasound images. The random collection of 16×16 dictionary atoms ($K = 256$) is presented in Figure 5a and the dictionary trained on the noisy image itself by overlapping patches is represented in Figure 5b. Where, every dictionary atom occupies a cell of 8×8 pixel ($N = 64$). We performed the tests on the three ultrasound reference images shown in Figures 6a, 7a and 9a. The KSVD algorithm was initialized with a trained dictionary and executed 180 iterations, as recommended in [29].

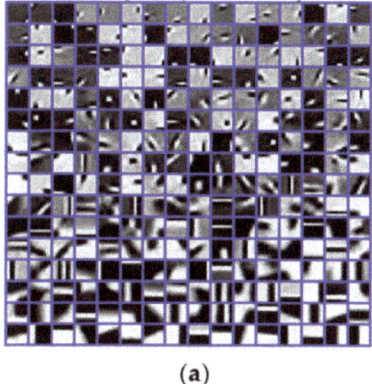 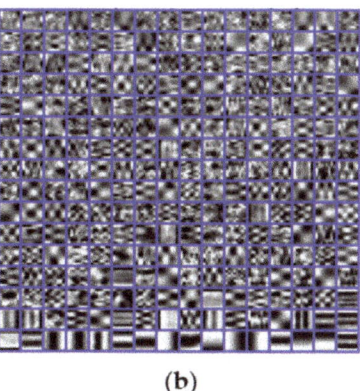

(a) (b)

Figure 5. The random collections of 16×16 atoms ($K = 256$) of trained dictionary from (**a**) a reference set of 3245 ultrasound images and (**b**) a noisy image.

The numerical evaluation was performed using PSNR and MSSIM (as discussed in Section 3.1) on the proposed algorithm and compared with the denoising methods Frost filter [10], Lee filter [11], 3×3 Weiner filter [13], Kuan filter [14], 3×3 median filter [15], SRAD filter [39], and wavelet filter [40].

Figure 6a, shows a right lobe liver image with size 256×256 pixels, where the lateral size is given by the *x*-axis, and the axial size is given by the *y*-axis. In this original image, we included a speckle noise parameter $\sigma = 10$ and the PSNR was calculated using Equation (12). It is clear that detailed information of the image is highly distorted, as shown in Figure 6b with a PSNR value of 28.148 dB. Figure 6c,d show the denoising results obtained by the proposed method using Dictionary 1 with a PSNR value of 35.033 dB and Dictionary 2 with a PSNR value of 35.537 dB. It is clear that the SR over learned dictionaries improves both edges and smooth features by eliminating the noise and reconstructs the image as much closer to the original image, as shown in Figure 6a.

Figure 7 shows the comparative experimental results obtained on real-time ultrasound images. For this experiment, we obtained a 256×256-pixel liver image of a healthy person with a PSNR value of 24.6271 dB. The radio frequency (RF) frames were obtained using a linear transducer with a frequency range of 8 MHz. This frequency range was selected because of its suitability for liver imaging, and we considered natural speckle noise for these experiments. The original speckled image was then denoised using the proposed algorithm with both dictionaries and also using conventional algorithms. To assess the speckle reduction, we selected two regions in of the speckled image. The two regions in the case of Figure 7a are displayed as a red square and a green square. The red one indicates the diaphragm of a liver and the green square shows the presence of an excessive noisy region observed from deeper tissue. The differences can be noticed from the filtered images in dashed red and the green square. Figure 7d–f show that detailed information lost by the blurring effect on the results obtained with Frost filter, median filter, and Kuan filter. In particular, the wavelet filter, Weiner filter, and the SRAD filter are not very effective in reducing speckle and perform poorly in retrieving sharp edge information, as can be seen in Figure 7g–i. Figure 7b shows the results for the proposed method using Dictionary 1 (PSNR = 30.3345 dB) and Figure 7c shows the results for the proposed method using Dictionary 2 (PSNR = 30.8073). It is clear that the image denoised using the proposed SR method reconstructed image very close to the original image. It can also be seen that the dictionary trained on the noisy image gives better results than using a set of multiple references images. The results of this comparative experiment show that the proposed algorithm not only reduces the speckle noise but also preserves the edge information. Table 2 shows the PSNR and MSSIM values to quantify the results numerically for noise parameter $\sigma = 15$.

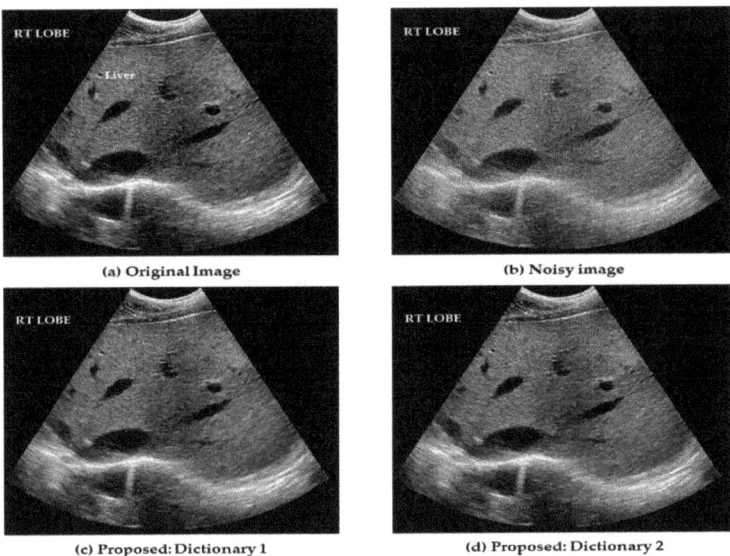

Figure 6. Reconstruction of liver right lobe images. (**a**) Original ultrasound image; (**b**) Speckled ultrasound image (PSNR = 28.148 dB); Images reconstructed using (**c**) Dictionary 1 (PSNR = 35.033 dB) and (**d**) Dictionary 2 (PSNR = 35.537 dB).

Table 2. PSNR and MSSIM for the ultrasound liver image for $\sigma = 15$.

Models	PSNR (dB)	MSSIM
Frost	28.966	0.822
Median	25.497	0.659
Wavelet	27.772	0.782
SRAD	28.766	0.813
Kuan	28.279	0.801
Weiner	29.218	0.834
Proposed: Dictionary 1	30.334	0.901
Proposed: Dictionary 2	30.807	0.926

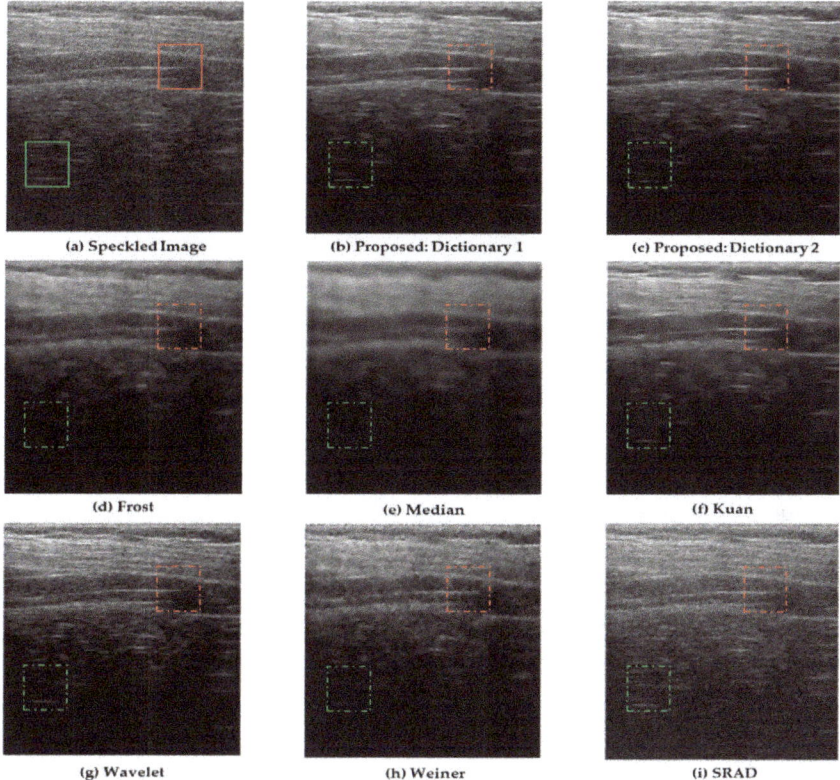

Figure 7. Despeckled results obtained for the ultrasound liver dataset using a linear transducer with a frequency of 8 MHz. The red and the green boxes highlight the differences observed from the noisy and filtered images. (**a**) Speckled image and results yielded by the proposed method using (**b**) Dictionary 1 and (**c**) Dictionary 2 as well as results using the (**d**) Frost; (**e**) median; (**f**) Kuan; (**g**) wavelet; (**h**) Weiner; and (**i**) SRAD filters.

Speckle is an arbitrary granular texture noise that degrades ultrasound image quality. This experiment was performed to evaluate different noise variances by comparing the PSNR obtained using the proposed algorithm and other despeckling algorithms. The simulated result using the noise levels 10, 15, 20, 25, and 30 are illustrated in Figure 8. The results clearly depict that, for different noise variances, the proposed algorithm gives the best PSNR value of all the algorithms on speckle reduction.

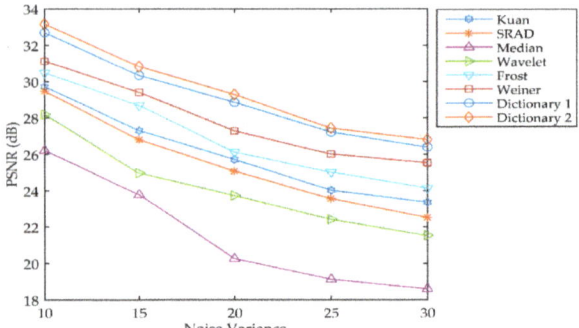

Figure 8. Comparison of PSNRs obtained by different methods. SRAD: speckle reducing anisotropic diffusion.

The experiments presented above were performed on ultrasound liver images, and the performance compared with conventional methods. However, our algorithm can also be utilized for a wide range of ultrasound images. To prove this, we conducted experiments on a real thrombus (blood clot) image with a left ventricular mass [56]. The visual assessment was performed using the proposed technique and the results compared to those obtained by various other algorithms. The reference image size was 256 × 256 pixels in order to fit our proposed model. The data were obtained from an open medical imaging dataset on GitHub [57]. The ultrasound image along with a marked note are shown in Figure 9a. The dashed white box in Figure 9b–j indicate regions of the ventricular mass. The thrombus data-set results presented in Figure 9h–j show that the wavelet, Weiner, and SRAD filters performed very poorly in noise reduction. The difference can be seen from the white note marked on the right atrium of the reference ultrasound image in Figure 9a. Figure 9e–g shows that Frost, median, and Kuan reduces speckle but tends to over-smooth the image, which leads to the loss of a distinctive feature of the unclear mass. Among all the methods, Figure 9c,d show good results for the SR-based on learned dictionaries 1 and 2. Several details are well preserved and the speckle noise is reduced efficiently. Figure 10 shows the zoomed sub-images of Figure 9 to observe a clear visualization of the despeckled images. The red box highlights the texture details in the noisy image and the filtered image for a comparative visual assessment. It can be noted that from the Frost, Median, and Kuan filtered data displayed in Figure 10d–f, an unclear mass (blood clot) and texture feature are blurred and over smoothed. Figure 10h,i show that the Weiner and SRAD filters are not much more effective on speckle reduction. These filters also greatly reduce the contrast, making images more indistinguishable from the background. This effect is especially noticeable in the case of the Wavelet filter as shown in Figure 10g. It was found that the anatomical structure was more clearly visible in Figure 10b,c obtained using the SR framework, where the speckle is reduced around the unclear mass without removing its features such as edges and texture. These results were comparatively better than those of Figure 10d–i of the standard despeckling methods. Thus, the proposed algorithm has various advantages for use in CAD systems based on image analysis, such as segmentation and edge detection. Future work will include extensive laboratory and clinical testing on diseased and healthy subjects for a more rigorous validation of the system. In conclusion, our approach reconstructs the detailed information in real ultrasound images, not only by preserving edge information but also by eliminating artifacts and reducing speckle noise.

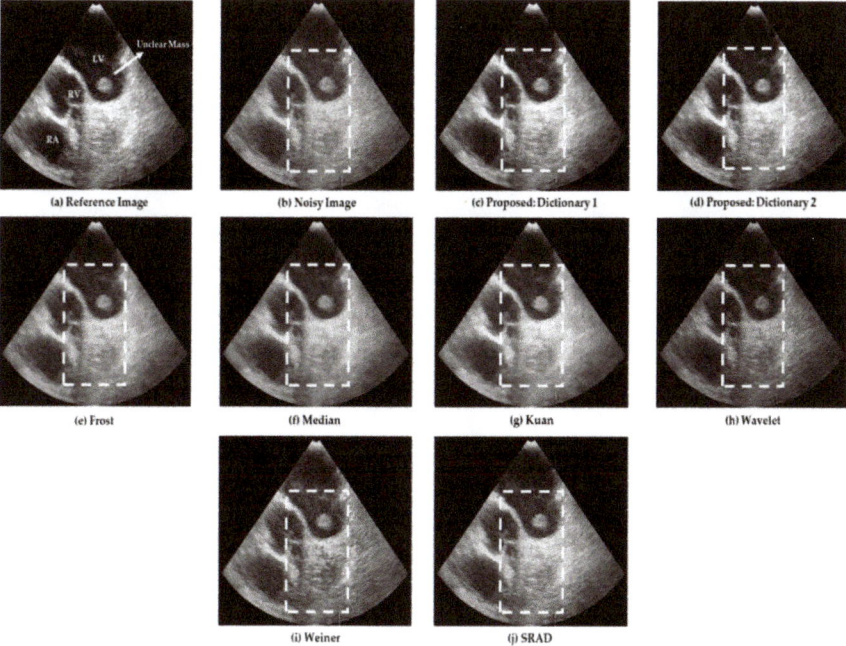

Figure 9. (**a**) Ultrasound image of the thrombus in the left ventricle. LV: left ventricle, RA: right atrium and RV: right ventricle and (**b**) noisy image. Despeckled ultrasound images of proposed method using (**c**) Dictionary 1 and (**d**) Dictionary 2. Results using the (**e**) Frost, (**f**) median, (**g**) Kuan, (**h**) wavelet, (**i**) Weiner, and (**j**) SRAD filters. The dashed white box indicates the region of image showing visual enhancement owing to despeckling.

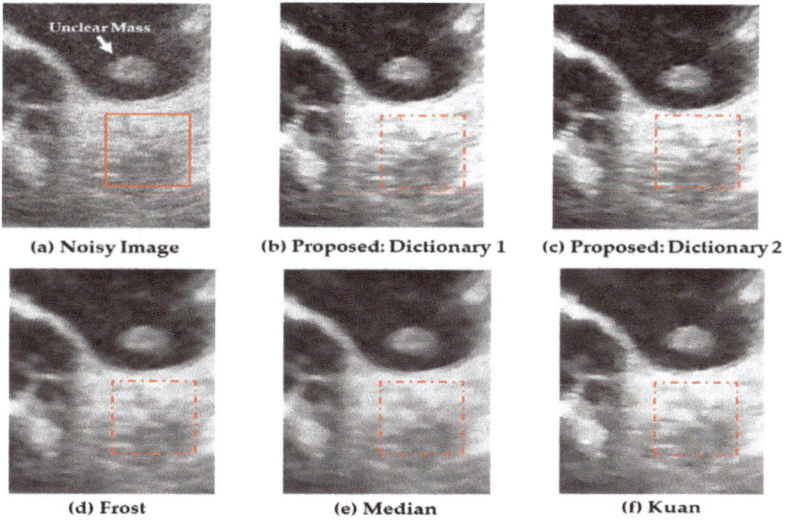

Figure 10. *Cont.*

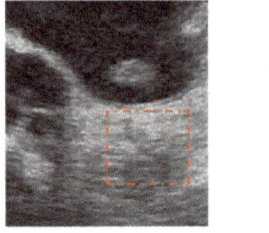 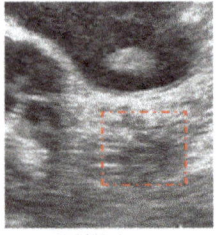 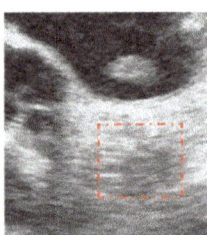

(g) Wavelet (h) Weiner (i) SRAD

Figure 10. (a) Zoomed sub-image of noisy thrombus ultrasound images. The red boxes highlight texture details of images for visual assessment. Results of proposed method using (b) Dictionary 1 and (c) Dictionary 2. Results using the (d) Frost; (e) median; (f) Kuan; (g) wavelet; (h) Weiner; and (i) SRAD filters.

5. Conclusions

In this paper, we presented a method that reconstructed ultrasound images by suppressing multiplicative speckle noise using the SR framework. The proposed method utilizes an enhanced homomorphic filter, TV regularization, and sparse prior over two learned dictionaries. In addition, the KSVD algorithm is used to train the two dictionaries—one trained with a set of reference ultrasound image patches and another trained with the speckled image patches. Both training options were tested with the synthetic images and various clinical ultrasound images. The experimental results obtained for different noise levels proved superior to those of other standard denoising methods. The results also show that the two modified dictionaries performed well with sparse and TV regularization terms. Overall, the proposed SR framework reconstructs the image signals by removing speckle noise while preserving the texture and yielding a smoother image than conventional methods without eliminating edges.

Author Contributions: M.Y.J. and H.N.L. formulated and designed the experiments; M.Y.J. performed the experiments and analyzed the data; M.Y.J. wrote the manuscript; H.N.L. made revisions to the manuscript and supervised the research work.

Funding: This work was supported by a National Research Foundation of Korea (NRF) grant provided by the Korean government (MSIP) [NRF-2018R1A2A1A19018665].

Conflicts of Interest: The authors declare no conflict of interest.

References

1. Szabo, T.L. *Diagnostic Ultrasound Imaging: Inside Out*; Academic Press Series in Biomedical Engineering; Elsevier Academic Press: New York, NY, USA, 2004; p. 549. ISBN 0-12-680145-2.
2. Martial, B.; Cachar, D. Acquire real-time RF digital ultrasound data from a commercial scanner. *Electron. J. Tech. Accoust.* **2007**, *3*, 16.
3. Lee, H.; Chen, Y.P.P. Image based computer aided diagnosis system for cancer detection. *Expert Syst. Appl.* **2015**, *42*, 5356–5365. [CrossRef]
4. Jabarulla, M.Y.; Lee, H.N. Computer aided diagnostic system for ultrasound liver images: A systematic review. *Optik* **2017**, *140*, 1114–1126. [CrossRef]
5. Zanotel, M.; Bednarova, I.; Londero, V.; Linda, A.; Lorenzon, M.; Girometti, R.; Zuiani, C. Automated breast ultrasound: Basic principles and emerging clinical applications. *Radiol. Med.* **2018**, *123*, 1–12. [CrossRef] [PubMed]
6. Acharya, U.R.; Koh, J.E.W.; Hagiwara, Y.; Tan, J.H.; Gertych, A.; Vijayananthan, A.; Yaakup, N.A.; Abdullah, B.J.J.; Fabell, M.K.B.M.; Yeong, C.H. Automated diagnosis of focal liver lesions using bidirectional empirical mode decomposition features. *Comput. Biol. Med.* **2017**, *94*, 11–18. [CrossRef] [PubMed]

7. Grazioli, L.; Ambrosini, R.; Frittoli, B.; Grazioli, M.; Morone, M. Primary benign liver lesions: Benign focal liver lesions can origin from all kind of liver cells: Hepatocytes, mesenchymal and cholangiocellular line. *Eur. J. Radiol.* **2017**, *26*, 378–398. [CrossRef] [PubMed]
8. Burckhart, C.B. Speckle in ultrasound B-mode scans. *IEEE Trans. Sonics Ultrason.* **1978**, *25*, 1–6. [CrossRef]
9. Narayanan, S. A view on despeckling in ultrasound imaging. *Int. J. Signal Process. Image Process. Pattern Recognit.* **2009**, *2*, 85–98.
10. Lopes, A.; Touzi, R.; Nezry, E. Adaptive Speckle Filters and Scene Heterogeneity. *IEEE Trans. Geosci. Remote Sens.* **1990**, *28*, 992–1000. [CrossRef]
11. Lee, J.S. Digital image enhancement and noise filtering by use of local statistics. *IEEE Trans. Pattern Anal. Mach. Intell.* **1980**, *2*, 165–168. [CrossRef] [PubMed]
12. Gonzalez, R.C.; Woods, R.E. *Digital Image Processing*, 3rd ed.; Pearson Education, Inc.: London, UK, 2008; ISBN 0-13-168728-x978-0-13-168728-8.
13. Goldstein, J.S.; Reed, I.S.; Scharf, L.L. A multistage representation of the Wiener filter based on orthogonal projections. *IEEE Trans. Inf. Theory* **1998**, *44*, 2943–2959. [CrossRef]
14. Kuan, D.T.; Sawchuk, A.A.; Strand, T.C.; Chavel, P. Adaptive noise smoothing filter for images with signal-dependent noise. *IEEE Trans. Pattern Anal. Mach. Intell.* **1985**, *7*, 165–177. [CrossRef] [PubMed]
15. Simon, P.; Patrick, H. Median Filtering in Constant Time. *IEEE Trans. Image Process.* **2007**, *16*, 2389–2394.
16. Achim, A.; Bezerianos, A.; Tsakalides, P. Novel Bayesian multiscale method for speckle removal in medical ultrasound images. *IEEE Trans. Med. Imaging* **2001**, *20*, 772–783. [CrossRef] [PubMed]
17. Chen, Z.J.; Chen, C.H.Y. Efficient statistical modeling of wavelet coefficients for image denoising. *Int. J. Wavelets Multiresolut. Inf. Process.* **2009**, *7*, 629–641. [CrossRef]
18. Vishwa, A.; Sharma, S. Modified method for denoising the ultrasound images by wavelet thresholding. *Int. J. Intell. Syst. Appl.* **2012**, *4*, 25. [CrossRef]
19. Shen, Y.; Liu, Q.; Lou, S.; Hou, Y.L. Wavelet-Based Total Variation and Nonlocal Similarity Model for Image Denoising. *IEEE Signal Process. Lett.* **2017**, *24*, 877–881. [CrossRef]
20. Donoho, D.L. De-noising by soft-thresholding. *IEEE Trans. Inf. Theory* **1995**, *41*, 613–627. [CrossRef]
21. Matsuyama, E.; Tsai, D.-Y.; Lee, Y.; Tsurumaki, M.; Takahashi, N.; Watanabe, H.; Chen, H.-M. A modified undecimated discrete wavelet transform based approach to mammographic image denoising. *J. Digit. Imaging* **2013**, *26*, 748–758. [CrossRef] [PubMed]
22. Kim, Y.S. Improvement of ultrasound image based on wavelet transform: Speckle reduction and edge enhancement. *SPIE Med. Imaging* **2005**, *5747*, 1085–1092.
23. Chambolle, A. An algorithm for total variation minimizations and applications. *J. Math. Imaging Vis.* **2004**, *10*, 89–97.
24. Rudin, L.I.; Osher, S.; Fatemi, E. Nonlinear total variation based noise removal algorithms. *Phys. D Nonlinear Phenom.* **1992**, *60*, 259–268. [CrossRef]
25. Perona, P.; Malik, J. Scale-space and edge detection using anisotropic diffusion. *IEEE Trans. Pattern Anal. Mach. Intell.* **1990**, *12*, 629–639. [CrossRef]
26. Chao, S.M.; Tsai, D.M. An improved anisotropic diffusion model for detail and edge-preserving smoothing. *Pattern Recognit. Lett.* **2010**, *31*, 2012–2023. [CrossRef]
27. Tschumperle, D.; Deriche, R. Vector-valued image regularization with PDEs: A common framework for different applications. *IEEE Trans. Pattern Anal. Mach. Intell.* **2005**, *27*, 506–517. [CrossRef] [PubMed]
28. Zhao, Y.; Yang, J. Hyperspectral image denoising via sparse representation and low-rank constraint. *IEEE Trans. Geosci. Remote Sens.* **2015**, *53*, 296–308. [CrossRef]
29. Elad, M.; Aharon, M. Image denoising via sparse and redundant representations over learned dictionaries in wavelet domain. *IEEE Trans. Image Process.* **2006**, *15*, 754–758. [CrossRef]
30. Deka, B.; Bora, P.K. Removal of correlated speckle noise using sparse and overcomplete representations. *Biomed. Signal Process. Control* **2013**, *8*, 520–533. [CrossRef]
31. Fan, J.; Wu, Y.; Li, M.; Liang, W.; Zhang, Q. SAR Image Registration Using Multiscale Image Patch Features with Sparse Representation. *Biomed. Signal Process. Control* **2017**, *10*, 1483–1493. [CrossRef]
32. Wright, J.; Ma, Y.; Mairal, J.; Sapiro, G.; Huang, T.S.; Yan, S. Sparse representation for computer vision and pattern recognition. *Proc. IEEE* **2010**, *98*, 1031–1044. [CrossRef]
33. Bruckstein, M.E.A.M.; Donoho, D.L.; Elad, M. From sparse solutions of systems of equations to sparse modeling of signals and images. *SIAM Rev.* **2009**, *51*, 34–81. [CrossRef]

34. Li, S.; Wang, G.; Zhao, X. Multiplicative noise removal via adaptive learned dictionaries and TV regularization. *Digit. Signal Process.* **2016**, *50*, 218–228.
35. Liu, K.; Tan, J.; Su, B. An Adaptive Image Denoising Model Based on Tikhonov and TV Regularizations. *Adv. Multimed.* **2014**, *2014*, 934834. [CrossRef]
36. Aharon, M.; Elad, M.; Bruckstein, A.M. The K-SVD: An algorithm for designing of overcomplete dictionaries for sparse representations. *IEEE Trans. Signal Process.* **2006**, *54*, 4311–4322. [CrossRef]
37. Tay, P.C.; Garson, C.D.; Acton, S.T.; Hossack, J.A. Ultrasound despeckling for contrast enhancement. *IEEE Trans. Image Process.* **2010**, *19*, 1847–1860. [CrossRef] [PubMed]
38. Joel, T.; Sivakumar, R. An extensive review on Despeckling of medical ultrasound images using various transformation techniques. *Appl. Acoust.* **2018**, *138*, 18–27. [CrossRef]
39. Youngjian, Y.; Acton, S.T. Speckle reducing anisotropic diffusion. *IEEE Trans. Image Process.* **2002**, *11*, 1260–1270. [CrossRef] [PubMed]
40. Hussain, S.A.; Gorashi, S.M. Image Denoising based on Spatial/Wavelet Filter using Hybrid Thresholding Function. *Int. J. Comput. Appl.* **2012**, *42*, 5–13.
41. Aubert, G.; Aujol, J.-F. A variational approach to removing multiplicative noise. *SIAM J. Appl. Math.* **2008**, *68*, 925–946. [CrossRef]
42. Buades, A.; Coll, B.; Morel, J.M. A review of image denoising algorithms, with a new one. *Multiscale Model. Simul.* **2005**, *4*, 490–530. [CrossRef]
43. Gilboa, S.O.G. Nonlocal operators with applications to image processing. *SIAM J. Multiscale Model. Simul.* **2008**, *7*, 1005–1028. [CrossRef]
44. Cai, T.T.; Wang, L. Orthogonal matching pursuit for sparse signal recovery with noise. *IEEE Trans. Inf. Theory* **2011**, *57*, 4680–4688. [CrossRef]
45. Deka, B.; Bora, P.K. Despeckling of medical ultrasound images using sparse representation. In Proceedings of the 2010 International Conference Signal Processing and Communications (SPCOM), Bangalore, India, 18–21 July 2010. ISSN 2165-0608.
46. Cobbold, R.S.C. *Foundations of Biomedical Ultrasound*; Oxford University Press: Oxford, UK, 2007.
47. Yahya, N.; Kamel, N.S.; Malik, A.S. Subspace-based technique for speckle noise reduction in ultrasound images. *Biomed. Eng. Online* **2014**, *13*, 154. [CrossRef] [PubMed]
48. Arsenault, H.H.; Levesque, M. Combined homomorphic and local-statistics processing for restoration of images degraded by signal-dependent noise. *Appl. Opt.* **1984**, *23*, 845–850. [CrossRef] [PubMed]
49. Xie, H.; Pierce, L.E.; Ulaby, F.T. Statistical properties of logarithmically transformed speckle. *IEEE Trans. Geosci. Remote Sens.* **2002**, *40*, 721–727. [CrossRef]
50. Candes, E.; Candes, E.; Romberg, J.; Romberg, J. *l1-Magic: Recovery of Sparse Signals via Convex Programming*; Caltech: Pasadena, CA, USA, 2005; pp. 1–19.
51. Afonso, M.V.; Bioucas-Dias, J.M.; Figueiredo, M.A.T. An augmented Lagrangian approach to the constrained optimization formulation of imaging inverse problems. *IEEE Trans. Image Process.* **2011**, *20*, 681–695. [CrossRef] [PubMed]
52. Davis, G.; Mallat, S.G.; Avellaneda, M. Adaptive greedy approximations. *Constr. Approx.* **1997**, *13*, 57–98. [CrossRef]
53. Xiang, F.; Wang, Z. Split Bregman iteration solution for sparse optimization in image restoration. *Optik* **2014**, *125*, 5635–5640. [CrossRef]
54. Wang, Z.; Bovik, A.; Sheikh, H.; Simoncelli, E. Image quality assessment: From error visibility to structural similarity. *IEEE Trans. Image Process.* **2004**, *4*, 600–612. [CrossRef]
55. Shepp, L.; Logan, F. The Fourier reconstruction of a head section. *IEEE Trans. Nucl. Sci.* **1974**, *21*, 21–43. [CrossRef]
56. Llach, F. Hypercoagulability, renal vein thrombosis, and other thrombotic complications of nephrotic syndrome. *Kidney Int.* **1985**, *3*, 429–439. [CrossRef]
57. GitHub. Available online: https://github.com/sfikas/medical-imaging-datasets (accessed on 1 May 2018).

© 2018 by the authors. Licensee MDPI, Basel, Switzerland. This article is an open access article distributed under the terms and conditions of the Creative Commons Attribution (CC BY) license (http://creativecommons.org/licenses/by/4.0/).

Article

A High-Efficiency Super-Resolution Reconstruction Method for Ultrasound Microvascular Imaging

Wei Guo [1,2], Yusheng Tong [3], Yurong Huang [1,2], Yuanyuan Wang [1,2,*] and Jinhua Yu [1,2]

1. Department of Electronic Engineering, Fudan University, Shanghai 200433, China; 13110720026@fudan.edu.cn (W.G.); 15210720136@fudan.edu.cn (Y.H.); jhyu@fudan.edu.cn (J.Y.)
2. Key Laboratory of Medical Imaging Computing and Computer Assisted Intervention of Shanghai, Fudan University, Shanghai 200032, China
3. Institute of Functional and Molecular Medical Imaging, Fudan University, Shanghai 200030, China; ystong16@fudan.edu.cn
* Correspondence: yywang@fudan.edu.cn

Received: 17 June 2018; Accepted: 5 July 2018; Published: 13 July 2018

Featured Application: This work makes real-time imaging of tumor microvessels more realistic.

Abstract: The emergence of super-resolution imaging makes it possible to display the microvasculatures clearly using ultrasound imaging, which is of great importance in the early diagnosis of cancer. At present, the super-resolution performance can only be achieved when the sampling signal is long enough (usually more than 10,000 frames). Thus, the imaging time resolution is not suitable for clinical use. In this paper, we proposed a novel super-resolution reconstruction method, which is proved to have a satisfactory resolution using shorter sampling signal sequences. In the microbubble localization step, the integrated form of the 2D Gaussian function is innovatively adopted for image deconvolution in our method, which enhances the accuracy of microbubble positioning. In the trajectory tracking step, for the first time the averaged shifted histogram technique is presented for the visualization, which greatly improves the precision of reconstruction. In vivo experiments on rabbits were conducted to verify the effectiveness of the proposed method. Compared to the conventional reconstruction method, our method significantly reduces the Full-Width-at-Half-Maximum (FWHM) by 50% using only 400-frame signals. Besides, there is no significant increase in the running time using the proposed method. Considering its imaging performance and used frame number, the conclusion can be drawn that the proposed method advances the application of super-resolution imaging to the clinical use with a much higher time resolution.

Keywords: ultrasound imaging; super-resolution; microbubble; reconstruction; time resolution

1. Introduction

Ultrasound has become a commonly used method of medical examination because of its fast, non-destructive and inexpensive characteristics. However, the main problem of ultrasound imaging lies in its low image quality. Compared to other modalities, such as the computed tomography (CT) and the magnetic resonance imaging (MRI), ultrasound imaging has a lower resolution. The reason is that the resolution of ultrasound imaging is limited by the emission wavelength [1]. Usually, the center frequency of the ultrasound is at the megahertz order, resulting in the wavelengths at the order of a hundred microns. With a resolution of this magnitude, it is not possible to observe the microvessels in microns, which are of great importance in the early diagnosis of cancer [2].

In 2016, M. Tanter et al. [3] proposed the epoch-making ultrasound imaging method called super-resolution ultrasound imaging. This method uses microbubbles as the resolution units and breaks through the resolution limitations of traditional ultrasound imaging. Through continuous plane wave transmissions, the complete trajectory of microbubbles can be recorded. After trajectory tracking, tiny blood vessels can be observed. This technique was tested on rat cerebral vascular and has been proved to be effective [3].

The concept of ultrasound super-resolution imaging originates from optical imaging [4–6], and both processes are very similar [7–9]. In ultrasound super-resolution imaging, the beamformed images corresponding to the plane wave transmissions are preprocessed firstly by the wall filter [10,11]. Then following are the two most important steps of the reconstruction, the microbubble localization and the trajectory tracking [7].

The sub-pixel localization of a single microbubble with an accuracy below the diffraction limit is the basis of super-resolution reconstruction methods [12]. Because a single microbubble can be treated as an incoherent point source in the beamformed image sequence, the result of fitting a point-spread-function (PSF) model to an image of a point-like emitter is an estimate of the microbubble position, its imaged size and intensity. It has been shown that the Gaussian function provides a very good approximation of the real PSF of an ultrasound scanner [13]. The advantages of Gaussian PSF models are their simplicity, robustness, and computational efficiency. However, due to the lack of adaptability, the original Gaussian PSF has a limited accuracy for positioning.

As for the trajectory tracking, the scatter plot is the simplest and most common visualization method [14], and does not always provide high-quality results. A simple binary image is created with the pixel intensity value set to one at locations corresponding to microbubble positions. All other pixel intensity values are set to zero. This method is fast but does not reflect the density of microbubbles.

Due to the above problems, the existing reconstruction methods can only get good super-resolution results when the collected signals are long enough (usually up to 10,000 frames) [15]. In this paper, we propose a novel super-resolution reconstruction method, in which brand-new microbubble localization and trajectory tracking techniques were presented for better reconstruction results with fewer sampling frames. Taking the characteristics of ultrasound images and the microbubble density into consideration, we innovatively adopted the integrated form of the 2D Gaussian function [16] and the averaged shifted histograms (ASH) [17] for the two steps respectively. The introduction of the methods helps enhance the quality of the reconstructed images significantly, and in turn improves the imaging time resolution. In order to prove the validity of the proposed method, we conducted in vivo experiments on the kidney of a rabbit.

The structure of the paper is arranged as follows: Materials and Methods section introduces the mathematical background of our proposed method. The setup of in vivo experiments is also given in this section. The corresponding results are shown in the Experimental Results section. Analysis and discussion of the resulting data are given in the Discussion section. Finally, we draw our short conclusion in the Conclusion section.

2. Materials and Methods

2.1. Super-Resolution Imaging Process

The process of super-resolution ultrasound imaging is quite like that of the single-molecule localization microscopy (SMLM) in optics [4,6], which is usually decomposed into four steps as shown in Figure 1.

In the SMLM, the first step is imaging light-activated fluorescent molecules that act as tiny, randomly distributed pinpricks of light. The use of low light intensities and the fact that the molecules' activation is inherently random ensures that only a sparse subset is turned on at any one time. Thus, these point-like light sources are separated by more than half a wavelength, so the image of each one (a blurred spot called the PSF) does not overlap with that of its neighbors. In ultrasound super-resolution imaging, the flowing microbubbles excited by the acoustic waves are similar to those molecules excited by the fluorescence. By controlling the injection concentration and injection volume, the microbubbles can reach the appropriate concentration. In this case, the microbubbles (or its group) do not overlap in vessels. The echoes enhanced by a microbubble group can be recorded by the ultrasound plane wave imaging, which uses the low emission energy and keeps the microbubbles unbroken [18–20]. After comparing sequential images, the locations of the few, well-separated, flowing microbubbles would be pinpointed.

The second step is to determine the exact position of each point-like source by finding the center of the PSF. This is possible for well-separated sources, because the shape of the PSF can be devised in advance. The resolution of the final imaging result depends largely on the design of the PSF. For this step, ultrasound imaging and fluorescence imaging are the same. We call this step microbubble localization in the following sections.

The third step is to repeat the illumination and detection steps many times. A different set of separated point-like sources is detected each time, until a sufficient density of source points has been obtained. In acoustics, microbubbles vibrate under the excitation of scanning ultrasound, so we do not need an extra operation like the illumination in optics.

In the final step, by marking the positions of all these point sources on a single meta-image, a super-resolved picture can be built up. The spatial resolution in this image can exceed the diffraction limit, because it is determined by the accuracy with which the position of each source can be estimated. The accuracy of the reconstructed super resolution image is closely related to the chosen visualization method. We call this step the trajectory tracking in the following sections.

According to the above analysis, two steps of the microbubble localization and trajectory tracking are most closely related to the quality of reconstructed images. Therefore, we improve the corresponding algorithms for these steps respectively in order to get more precise super-resolution reconstruction images in this paper.

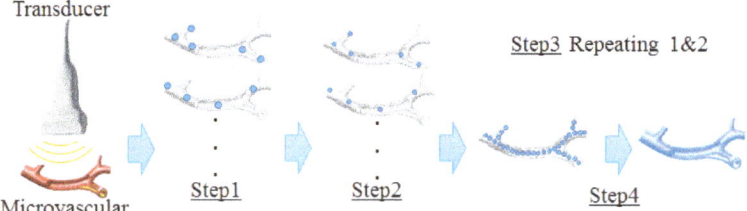

Figure 1. Super-resolution imaging process: Step 1. Scanning and comparing, Step 2. Microbubble localization, Step 3. Repeating 1 & 2 for each frame, Step 4. Trajectory tracking.

2.2. Microbubble Localization

2.2.1. PSF Models

The impulse response of an ultrasound scanner to a point-like source is described by the PSF. A common approximation of the real PSF is a symmetric two-dimensional Gaussian function given by the formula [13]:

$$PSF_G(x,y|\theta) = \frac{\theta_N}{2\pi\theta_\sigma^2} exp\left(-\frac{(x-\theta_x)^2 + (y-\theta_y)^2}{2\theta_\sigma^2}\right) + \theta_b, \qquad (1)$$

where $PSF_G(x,y|\theta)$ gives the expected microbubble count at the integer pixel position (x,y) for the parameters $\theta = [\theta_x; \theta_y; \theta_\sigma; \theta_N; \theta_b]$. The entries of the vector θ are as follows: θ_x and θ_y are the sub-pixel microbubble (group) coordinates, θ_σ is the imaged size of the microbubble (group), θ_N corresponds to the total number of the microbubbles at this position, and θ_b corresponds to the background noise level.

Though being simple, the original two-dimensional Gaussian function omits the prior knowledge about the imaging priori knowledge. The integrated form of a symmetric two-dimensional Gaussian function [21] is used in our method to help take into account the discrete nature of pixels presented in ultrasound images. Assuming a uniform distribution of pixels with the unit size, a single microbubble intensity profile can be expressed as:

$$PSF_G(x,y|\theta) = \theta_N E_x E_y + \theta_b, \tag{2}$$

where $PSF_{IG}(x,y|\theta)$ gives the expected microbubble count at the integer pixel position (x,y) for the parameters $\theta = [\theta_x; \theta_y; \theta_\sigma; \theta_N; \theta_b]$ and

$$E_x = \frac{1}{2}\mathrm{erf}\left(\frac{x - \theta_x + \frac{1}{2}}{\sqrt{2}\theta_\sigma}\right) - \frac{1}{2}\mathrm{erf}\left(\frac{x - \theta_x - \frac{1}{2}}{\sqrt{2}\theta_\sigma}\right), \tag{3}$$

$$E_y = \frac{1}{2}\mathrm{erf}\left(\frac{y - \theta_y + \frac{1}{2}}{\sqrt{2}\theta_\sigma}\right) - \frac{1}{2}\mathrm{erf}\left(\frac{y - \theta_y + \frac{1}{2}}{\sqrt{2}\theta_\sigma}\right). \tag{4}$$

2.2.2. Data Approximation

Given the approximate position of a microbubble (group) $(\tilde{x}_p, \tilde{y}_p)$ and a user-specified fitting radius $r > 0$, we define $D = \{-r, \ldots, r\} \times \{-r, \ldots, r\}$ as a set of integer (x,y) coordinates, and $\tilde{I}(x,y) = I(x + \tilde{x}_p, y + \tilde{y}_p)$ as intensity values of an $l \times l$ sub-image centered at the point $(\tilde{x}_p, \tilde{y}_p)$ of the raw input image I, where $l = 2r + 1$ is the size of the subimage. The desired sub-pixel coordinates of the microbubbles are obtained as $\hat{x}_p = \hat{x}_0 + \tilde{x}_p$ and $\hat{y}_p = \hat{y}_0 + \tilde{y}_p$, where $\hat{x}_0$ and $\hat{y}_0$ define the sub-pixel refinements of the coordinates obtained by the data approximation.

To approximate the data with a PSF, least-squares methods are employed to minimize the sum of (weighted) squared residuals defined by the data approximation method [22]

$$X^2(\theta|D) = \sum_{x,y \in D} w\left(\tilde{I}(x,y) - PSF(x,y|\theta)\right)^2. \tag{5}$$

Here the residual value for the (x,y) data point is defined as the difference between the observed image intensity $\tilde{I}(x,y)$ and the value approximated by the $PSF(x,y|\theta)$, where θ are the PSF parameters. The residual value can be further weighted by $w = 1$, making all measurements equally significant, or weighted by $w = 1/\tilde{I}(x,y)$, which takes into account the uncertainty in the number of detected microbubbles [23].

The search for parameters $\hat{\theta}$ which minimize $X^2(\theta|D)$, leads to an optimization problem formulated as [24]

$$\hat{\theta} = \arg\min_\theta X^2(\theta|D), \tag{6}$$

which we solve by the Levenberg-Marquardt algorithm as implemented in the Apache Commons Math library [25]. The starting point for the optimization process is computed from the data as the difference between the maximum and the minimum intensity values for the microbubble intensity θ_N, and as the minimum intensity value for the background offset θ_b. Users have to choose the starting point for the approximate microbubble (group) width θ_σ. The sub-pixel refinement of the coordinates is obtained as $\hat{x}_0 = \hat{\theta}_x$ and $\hat{y}_0 = \hat{\theta}_y$, where $\hat{\theta} = [\hat{\theta}_x, \hat{\theta}_y, \ldots]$.

For constraining parameters of PSF models, the Levenberg-Marquardt algorithm used above searches for values of the parameters θ over an infinite interval. The optimization process can therefore converge to a solution with negative values which is impossible for variables corresponding to the image intensity or to the standard deviation of a Gaussian PSF. We therefore limit the interval of possible values by transforming the relevant parameters and using PSF $(x, y|\tilde{\theta})$ in Equation (2) instead of PSF $(x, y|\theta)$. The transformation for a 2D Gaussian PSF model is $\tilde{\theta} = [\theta_x, \theta_y, \theta_\sigma^2, \theta_N^2, \theta_b^2]$.

The optimization process is still unconstrained but will result in positive PSF parameters. Such a transformation also improves the stability of the fit.

2.2.3. Localization Uncertainty

Localization uncertainty works as an evaluation metric measuring the accuracy of microbubble positioning. Let σ be the standard deviation of a fitted Gaussian PSF, a is the super-resolution reconstructed pixel size, N is the number of microbubbles detected for a given location, and b is the background signal level in microbubbles calculated as the standard deviation of the residuals between the raw data and our fitted PSF model [26]. The uncertainty in the lateral position of a microbubble can be approximated by the following formula [27]

$$\langle \Delta x^2 \rangle = \frac{\sigma^2 + a^2/12}{N} + \frac{8\pi\sigma^4 b^2}{a^2 N^2}. \tag{7}$$

As for the integrated Gaussian function, this formula still holds. Meanwhile, the σ is reduced and the localization accuracy increased.

2.3. Trajectory Tracking

Visualization (rendering) of the processed sequence data involves the creation of a new super-resolution image based on the coordinates of the localized microbubbles. The original scatter plot method [14] is a typical all-or-none tracking method. When a microbubble appears at a certain place, it is simply recorded and reflected with the constant gray level in the reconstruction result. It cannot display the density of microbubbles and the exact shape of the blood flow vessels. Thus, it has a negative impact on the imaging resolution. We solved this problem based on the histograms.

Histograms are often used to estimate the density of data by counting the number of observations that fall into each bin [14]. In our case, a two-dimensional histogram of microbubble positions can be created with the bin size corresponding to the pixel size of the final super-resolution image. Thus, for every localized microbubble (group), the bin value (i.e., the image brightness) at the corresponding microbubble positions is incremented by one. For a random sample of size h, the classic histogram takes the value

$$\hat{f}(x) = \frac{\text{bin count}}{h}. \tag{8}$$

The histogram visualization optionally supports "jittering" of the microbubbles between frames. When enabled, a random number drawn from the normal distribution, with a standard deviation equal to the computed (or user-specified) localization uncertainty, is added to the coordinates of every microbubble position before creating the histogram. This step is applied several times and all generated histograms are averaged together. As the number of jitters increases, the final image approaches the result of the Gaussian rendering, whose result is often considered as the golden standard in the SMLM. It can reflect the density of microbubbles and depict an accurate vascular boundary. For a small number of jitters, the histogram visualization is much faster than the Gaussian rendering but the resulting images may appear noisy. To solve this problem, we introduced an improved visualization algorithm.

This visualization algorithm uses a density estimation approach based on ASH [16]. The averaged shifted histogram or ASH is a nonparametric probability density estimator derived from a collection of histograms [28], as shown in Figure 2. In the one-dimensional case, this method works by averaging n histograms with the same bin width w, but with the origin of each histogram shifted by $\frac{w}{n}$ from the previous histogram. In the multidimensional case, there are n^d multidimensional histograms averaged in total, i.e., for n shifts in each of d dimensions. In our implementation, the width of the histogram bin is determined as $w = na$, where a is the pixel size of the super-resolution image. The number of shifts n

in the lateral and axial directions can be specified independently. A simple calculation shows that the ordinary ASH is given by

$$\hat{f}(x) = \frac{1}{n} \sum_{i=1-n}^{n-1} \frac{(n-|i|) \text{bin count}_i}{h}. \tag{9}$$

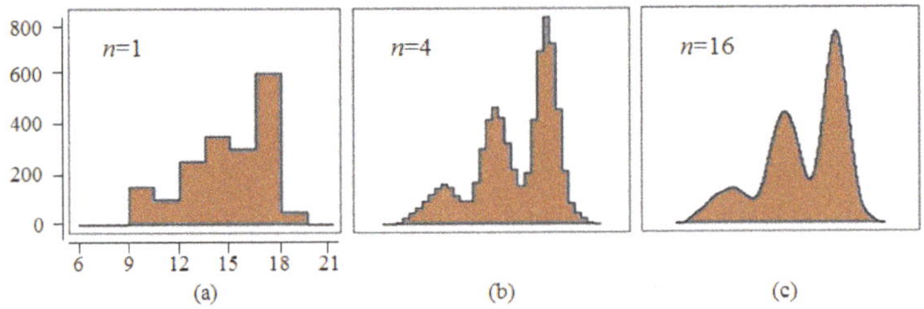

Figure 2. A simple example of the ASH method for the step count in one day: (**a**) a histogram, (**b**) an ASH estimate of 4 shifted histograms, and (**c**) an ASH estimate of 16 shifted histograms.

Theoretically, the time complexity of the proposed method is $O(N)$, where N is the number of microbubbles to visualize. This complexity level is close to the conventional scatter plot approach. However, the real speed of our visualization method is also influenced by the number of histograms to average, which makes it a bit slower than the scatter plot. From the perspective of the reconstruction performance, the proposed method based on the ASH approach provides similar results as Gaussian rendering with a constant localization uncertainty. However, the ASH approach is orders of magnitude faster than Gaussian rendering.

2.4. In Vivo Experiment Setups

2.4.1. In Vivo Ultrafast Imaging

The proposed method was evaluated using a New Zealand white rabbit model. Normal vasculature was imaged in the kidney of a healthy rabbit using the pulse inversion (PI) technique. The rabbits weighed 3.5–4.5 kg. Before the experiment, we carried out depilation on the rabbit's abdomen. The in vivo scans were performed using the Verasonics ultrasound system (Verasonics, Kirkland, WA, USA) with channel-domain data acquisition capabilities. A 6.3 MHz, 128-element linear array transducer (L11-4v) (Verasonics, Kirkland, WA, USA) with a pitch of 0.30 mm was used and the sampling frequency was set to 36 MHz. One and a half cycle Gaussian envelope-modulated pulses with a carrier frequency of 4.5 MHz were used in order to capture the harmonic components. Entire images were acquired using plane-wave acquisitions at a frame rate of 750 Hz (pulse repetition frequency of 1500 Hz) and low mechanical index of 0.05. Short cine loop of 400 frames were acquired for the rabbit model. Sonovue (Bracco, Milan, Italy), a clinically approved commercial ultrasound contrast agent [29], was used in our experiment. The microbubbles were injected via the ear vein of the rabbit in bolus injections of 1 mL with concentration of 10 µL/Kg and flushed with an additional 1 mL of saline. All experimental protocols were under the approval of the ethics and academic committee with the number (FD-ZS2016-082).

2.4.2. Radio-Frequency (RF) Data Processing and Wall Filtering

The RF echoes were first summed in order to separate the signal nonlinear component from the linear clutter. Then the second harmonics were filtered by a finite impulse response (FIR) bandpass filter with a 4 to 11 MHz passband and then beamformed with a delay-and-sum (DAS) beamformer. A wall filter based on the eigen-decomposition [30] was then used to separate the echoes of flowing microbubbles from the weak harmonic tissue clutter and stationary microbubbles.

Following References [10,30], the cut-off eigen-value was chosen as 0.2·max_eigen value. This is the lowest cut-off possible in order to include slow flowing microbubbles while removing clutter artifacts.

2.4.3. Methods for Comparison

Apart from the method in Reference [3] (Gaussian PSF deconvolution + scatter plot) and our proposed reconstruction method, a method based on super-resolution optical fluctuation imaging (SOFI) [31] is included for comparison too. This method calculates high-order moments for the microbubble localization and uses the power Doppler integral for the trajectory tracking. Generally, the SOFI method requires a short sampling sequence and short reconstruction time. In the meantime, the accuracy of its reconstruction results is not so good. In this method, the 4th order central moments of the RF signal were calculated for a single frame time-lag. In our method, a is set to 2 μm and n is set to 10. The super-resolution reconstruction algorithms are implemented in Matlab® and Image J® [32] with the Thunderstorm® plugin (version 1.3) [33] on a standard desktop PC with an Intel® Xeon® CPU E5-2637 v2 3.50 GHz with 64 GB of RAM.

2.4.4. Parameters Measurement

We use the Full-Width at Half-Maximum (FWHM), defined as the −6 dB bandwidth for the mainlobe, and Peak Sidelobe Level (PSL), defined as the peak value of the first sidelobe, to quantify the performance of different methods for imaging microvasculars. The former corresponds to the lateral resolution and the latter corresponds to the sidelobe level. To compare the efficiency of different methods, we also record their runtime in seconds.

3. Experimental Results

In this section, the results of three different super-resolution imaging schemes are shown for comparison using the same ultrafast plane wave dataset.

3.1. Preprocess

In order to demonstrate the whole process of super-resolution imaging more clearly, we show the results of each step from the start in Figure 3. Figure 3a shows the imaging result after the DAS. As seen, the echoes of the background tissues are strong, and the microbubbles cannot be observed. After the wall filtering, the signal of the contrast agents has been highlighted in Figure 3b, but the structure of the microvessels still cannot be seen at this time. After super-resolution reconstruction using the Gaussian PSF deconvolution and scatter plot is taken, the microvascular structure is finally clearly displayed, as shown in Figure 3c.

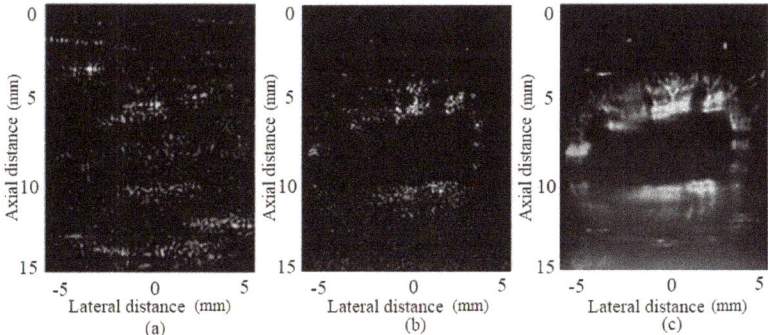

Figure 3. The imaging results of rabbit kidney using: (**a**) DAS, (**b**) DAS + wall filter, and (**c**) DAS + wall filter + Gaussian PSF deconvolution + scatter plot respectively. All images are shown with a dynamic range of 50 dB.

3.2. Results of Different Methods

Figure 4 shows the super-resolution imaging results of three reconstruction algorithms for the same dataset. Because only the echoes of 400 sampling frames are used in the imaging process, the resolution of the image reconstructed by Gaussian PSF deconvolution and scatter plot is limited. The high-order moment calculation and power Doppler integration method performs better in suppressing the background noise but the resolution of the reconstructed image is worse and not satisfactory. As for the vessels in the right bottom corner pointed by the blue arrows, this method fails to figure them out. The proposed method presents the finest microvascular structure and also has a certain inhibitory effect on the artefacts. For the artefacts in the left bottom corner pointed by the yellow arrows, the proposed method has the best suppressing performance. In order to observe the details of the image more clearly, we selected and enlarged a specific area of the image, and the results are shown in Figure 5. The analysis object is the vessels at the top right corner marked by the orange rectangles in Figure 4. As the results show, the proposed method is the only one that can clearly display the microvascular structure, which suggests that our method obtains the best resolution.

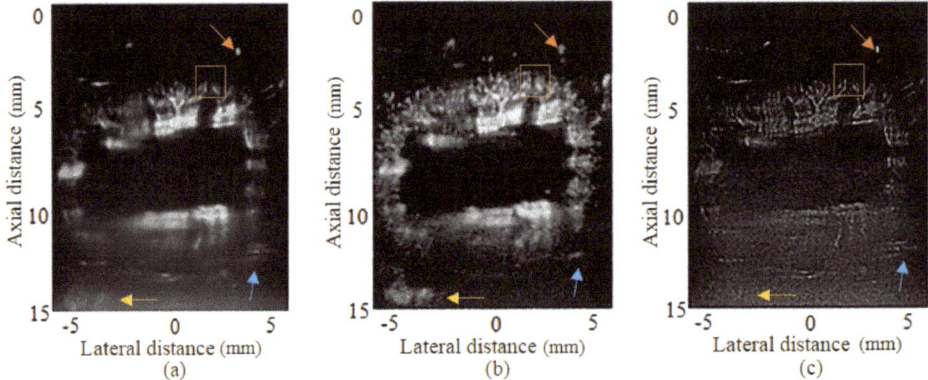

Figure 4. The super-resolution construction imaging results of rabbit kidney using: (**a**) Gaussian PSF deconvolution + scatter plot, (**b**) High-order moment calculation + power Doppler integration, and (**c**) Integral form of Gaussian PSF deconvolution + ASH respectively. All images are shown with a dynamic range of 50 dB. The orange, yellow and blue arrows point to the point target, artefact and the vascular details respectively. The orange box marks the area we choose to enlarge.

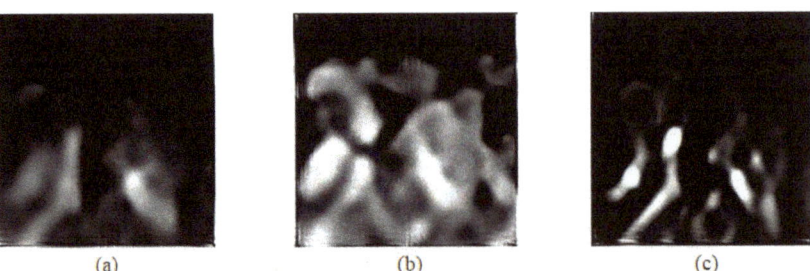

Figure 5. Locally magnified (10 times) microvascular images of rabbit kidney using: (**a**) Gaussian PSF deconvolution + scatter plot, (**b**) High-order moment calculation + power Doppler integration, and (**c**) Integral form of Gaussian PSF deconvolution + ASH respectively. All images are shown with a dynamic range of 50 dB.

To further compare the resolution performance, Figure 6 shows the lateral variation of the point target on the right top corner in Figure 4, which is pointed by the orange arrows. As seen, our proposed method performs best in the lateral resolution while the moment calculation method performs worst.

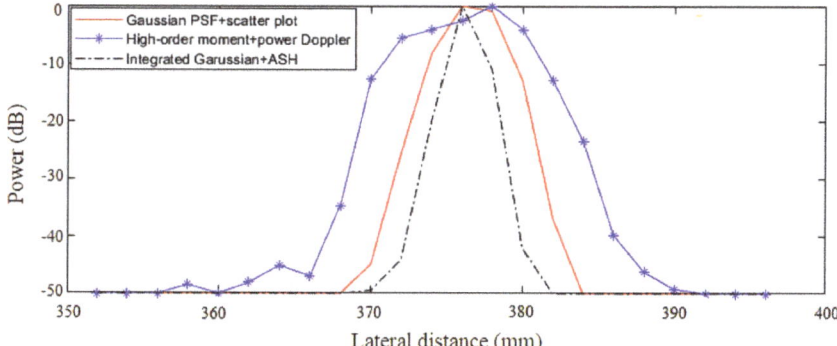

Figure 6. The lateral variation of different reconstruction methods for the point target at the right top corner.

Table 1 gives the statistical results for the resolution and running time of different methods. The result is the same as that of qualitative observation. The best FWHM and PSL indices are obtained by the proposed method. In terms of FWHM, the value of the proposed method is reduced by half in comparison to the original Gaussian PSF method. Meanwhile, the reduction of the PSL exceeds 5 dB. Though our method is not the fastest, its running time is still in the acceptable range. Considering the short sequence that was used, the improvement in the time resolution by our method is obvious.

Table 1. Full-Width at Half-Maximum (FWHM), Peak Side Lobe (PSL) and the run time of different reconstruction methods.

Method	FWHM (µm)	PSL (dB)	Run Time (s)
Gaussian PSF deconvolution + scatter plot	7.72	−37.22	14.20
High-order moment + power Doppler integration	13.44	−26.46	9.27
Integral form of Gaussian PSF deconvolution + ASH	3.88	−42.35	17.11

4. Discussion

In this paper, we propose a new super-resolution construction method for ultrasound imaging. Compared to the existing methods [3,31,34–36], the innovations of the proposed method lie in two aspects. In the microbubble localization step, we adopt a new Gaussian PSF to measure the accurate location of the microbubbles. This kind of Gaussian PSF takes the discreteness of the ultrasonic imaging coordinates into consideration and obtains more accurate positioning result. In the trajectory tracking step, we introduce the ASH method. The ASH enjoys several advantages compared with the conventional histogram method: better visual interpretation, better approximation, and nearly the same computational efficiency. According to the comparison of the methods in Results, our method has two main advantages over the other super-resolution construction methods:

First, the proposed method acquires the best resolution. This is because our method takes into account the characteristics of ultrasonic imaging when calculating the PSF. Besides, a more accurate visualization method is adopted.

Second, our approach has better real-time performance. Basically, the frames needed for the reconstruction are decreased in our method, which in turn increases the frame rate.

Besides, our algorithm still keeps a low amount of computation. Thus, our proposed method makes a step forward to the real-time super-resolution imaging.

At present, our improved method has achieved good results in displaying the rabbit kidney microvessels. In order to further verify the effectiveness of the proposed method, we need to carry out more in vivo animal experiments and clinical experiments in the future.

5. Conclusions

This paper aims to improve the original super-resolution ultrasound reconstruction imaging method, and achieve more accurate imaging results with shorter acquisition dataset. To this end, a novel super-resolution reconstruction method is put forward. The integrated form of a symmetric two-dimensional Gaussian function is introduced to locate the microbubble center after wall filtering. Then, the ASH is employed to visualize the microvessels in the reconstruction result. In vivo experimental results demonstrate that our proposed super-resolution ultrasound reconstruction method can obtain better performance in comparison with the original method in Reference [3] and other methods using a shorter dataset. Although our method shows certain potential in the existing experiments, a clinical experiment is needed in the future.

Authors Contributions

Formal Analysis & Conceptualization, W.G.; Validation, W.G., Y.T. and Y.H.; Writing-Original Draft Preparation, W.G.; Writing-Review & Editing, Y.W. and J.Y.; Supervision, Y.W. and J.Y.; Funding Acquisition, Y.W.

Funding: This work is supported by the National Natural Science Foundation of China (61771143 and 61471125).

Conflicts of Interest: The authors declare no conflict of interest.

References

1. Cootney, R.W. Ultrasound imaging: Principles and applications in rodent research. *Ilar J.* **2001**, *42*, 233–247. [CrossRef]
2. Jain, R.K. Normalization of tumor vasculature: An emerging concept in antiangiogenic therapy. *Science* **2005**, *307*, 58–62. [CrossRef] [PubMed]
3. Errico, C.; Pierre, J.; Pezet, S.; Desailly, Y.; Lenkei, Z.; Couture, O.; Tanter, M. Ultrafast ultrasound localization microscopy for deep super-resolution vascular imaging. *Nature* **2015**, *527*, 499–502. [CrossRef] [PubMed]
4. Rust, M.J.; Bates, M.; Zhuang, X. Sub-diffraction-limit imaging by stochastic optical reconstruction microscopy (STORM). *Nat. Method* **2006**, *3*, 793–796. [CrossRef] [PubMed]
5. Betzig, E.; Patterson, G.H.; Sougrat, R.; Lindwasser, O.W.; Olenych, S.; Bonifacino, J.S.; Davidson, M.W.; Lippincott-Schwartz, J.; Hess, H.F. Imaging intracellular fluorescent proteins at nanometer resolution. *Science* **2006**, *313*, 1642–1645. [CrossRef] [PubMed]
6. Hess, S.T.; Girirajan, T.P.K.; Mason, M.D. Ultra-high resolution imaging by fluorescence photoactivation localization microscopy. *Biophys. J.* **2006**, *91*, 4258–4272. [CrossRef] [PubMed]
7. Errico, C.; Couture, O.; Tanter, M. Ultrafast ultrasound localization microscopy. *J. Acoust. Soc. Am.* **2017**, *141*, 3951. [CrossRef]
8. Ghosh, D.; Xiong, F.; Sirsi, S.R.; Mattrey, R.; Brekken, R.; Kim, J.W.; Hoyt, K. Monitoring early tumor response to vascular targeted therapy using super-resolution ultrasound imaging. In Proceedings of the 2017 IEEE International Ultrasonics Symposium (IUS), Washington, DC, USA, 6–9 September 2017.
9. Lin, F.; Tsuruta, J.K.; Rojas, J.D.; Dayton, P.A. Optimizing sensitivity of ultrasound contrast-enhanced super-resolution imaging by tailoring size distribution of microbubble contrast agent. *Ultrasound Med. Biol.* **2017**, *43*, 2488–2493. [CrossRef] [PubMed]
10. Yu, A.; Lovstakken, L. Eigen-based clutter filter design for ultrasound color flow imaging: A review. *IEEE Trans. Ultrason. Ferroelectr. Freq. Control* **2010**, *57*, 1096–1111. [CrossRef] [PubMed]
11. Gallippi, C.M.; Trahey, G.E. Adaptive clutter filtering via blind source separation for two-dimensional ultrasonic blood velocity measurement. *Ultrasonic Imaging* **2002**, *24*, 193–214. [CrossRef] [PubMed]

12. Thompson, R.E.; Larson, D.R.; Webb, W.W. Precise nanometer localization analysis for individual fluorescent probes. *Biophys. J.* **2002**, *82*, 2775–2783. [CrossRef]
13. Stallinga, S.; Rieger, B. Accuracy of the Gaussian point spread function model in 2D localization microscopy. *Opt. Express* **2010**, *18*, 24461–24476. [CrossRef] [PubMed]
14. Baddeley, D.; Cannell, M.B.; Soeller, C. Visualization of localization microscopy data. *Microsc. Microanal.* **2010**, *16*, 64–72. [CrossRef] [PubMed]
15. Rix, A.; Lederle, W.; Theek, B.; Lammers, T.; Moonen, C.; Schmitz, G.; Kiessling, F. Advanced ultrasound technologies for diagnosis and therapy. *J. Nucl. Med.* **2018**, *59*, 740–746. [CrossRef] [PubMed]
16. Huang, F.; Schwartz, S.L.; Byars, J.M.; Lidke, K.A. Simultaneous multiple emitter fitting for single molecule super-resolution imaging. *Biomed. Opt. Express* **2011**, *2*, 1377–1393. [CrossRef] [PubMed]
17. Scott, D.W. Averaged shifted histograms: Effective nonparametric density estimators in several dimensions. *Ann. Stat.* **1985**, *13*, 1024–1040. [CrossRef]
18. Tanter, M.; Fink, M. Ultrafast imaging in biomedical ultrasound. *IEEE Trans. Ultrason. Ferroelectr. Freq. Control* **2014**, *61*, 102–119. [CrossRef] [PubMed]
19. Tiran, E.; Deffieux, T.; Correia, M.; Maresca, D.; Osmanski, B.F.; Sieu, L.A.; Bergel, A.; Cohen, I.; Pernot, M.; Tanter, M. Multiplane wave imaging increases signal-to-noise ratio in ultrafast ultrasound imaging. *Phys. Med. Biol.* **2015**, *60*, 8549–8566. [CrossRef] [PubMed]
20. Cox, B.; Beard, P. Imaging techniques: Super-resolution ultrasound. *Nature* **2015**, *527*, 451–452. [CrossRef] [PubMed]
21. Smith, C.S.; Joseph, N.; Rieger, B.; Lidke, K.A. Fast, single-molecule localization that achieves theoretically minimum uncertainty. *Nat. Method* **2010**, *7*, 373–375. [CrossRef] [PubMed]
22. Kendall, M.; Stuart, A. *The Advanced Theory of Statistics*, 2nd ed.; Charles Griffin: London, UK, 1946; pp. 29–48. ISBN 9780340814932.
23. Mortensen, K.I.; Churchman, L.S.; Spudich, J.A.; Flyvbjerg, H. Optimized localization analysis for single-molecule tracking and super-resolution microscopy. *Nat. Method* **2010**, *7*, 377–381. [CrossRef] [PubMed]
24. Bevington, P.R.; Robinson, D.K.; Blair, J.M.; Mallinckrodt, A.J.; McKay, S. Data reduction and error analysis for the physical sciences. *Comput. Phys.* **1993**, *7*, 415–416. [CrossRef]
25. Math, C. The Apache Commons Mathematics Library, Version 3.2. Available online: http://commons.apache.org/math/ (accessed on 4 February 2018).
26. Křížek, P.; Raška, I.; Hagen, G.M. Minimizing detection errors in single molecule localization microscopy. *Opt. Express* **2011**, *19*, 3226–3235. [CrossRef] [PubMed]
27. Quan, T.; Zeng, S.; Huang, Z.L. Localization capability and limitation of electron-multiplying charge-coupled, scientific complementary metal-oxide semiconductor, and charge-coupled devices for superresolution imaging. *J. Biomed. Opt.* **2010**, *15*, 066005. [CrossRef] [PubMed]
28. Scott, D.W. *Multivariate Density Estimation: Theory, Practice, and Visualization*, 2nd ed.; John Wiley & Sons: Hoboken, NJ, USA, 2015; pp. 242–245. ISBN 9780471697558.
29. Schneider, M. Sonovue, a new ultrasound contrast agent. *Eur. Radiol.* **1999**, *9*, S347–S348. [CrossRef] [PubMed]
30. Demené, C.; Deffieux, T.; Pernot, M.; Osmanski, B.F.; Biran, V.; Gennisson, J.L.; Sieu, L.; Bergel, A.; Franqui, S.; Correas, J.M.; et al. Spatiotemporal clutter filtering of ultrafast ultrasound data highly increases doppler and fultrasound sensitivity. *IEEE Trans. Med. Imaging* **2015**, *34*, 2271–2285. [CrossRef] [PubMed]
31. Bar-Zion, A.; Tremblay-Darveau, C.; Solomon, O.; Adam, D.; Eldar, Y.C. Fast vascular ultrasound imaging with enhanced spatial resolution and background rejection. *IEEE Trans. Med. Imaging* **2017**, *36*, 169–180. [CrossRef] [PubMed]
32. Schneider, C.A.; Rasband, W.S.; Eliceiri, K.W. NIH Image to ImageJ: 25 Years of image analysis. *Nat. Method* **2012**, *9*, 671–675. [CrossRef]
33. Ovesný, M.; Křížek, P.; Borkovec, J.; Švindrych, Z.; Hagen, G.M. ThunderSTORM: A comprehensive ImageJ plug-in for PALM and STORM data analysis and super-resolution imaging. *Bioinformatics* **2014**, *30*, 2389–2390. [CrossRef] [PubMed]
34. Viessmann, O.M.; Eckersley, R.J.; Christensenjeffries, K.; Tang, M.X.; Dunsby, C. Acoustic super-resolution with ultrasound and microbubbles. *Phys. Med. Biol.* **2013**, *58*, 6447–6458. [CrossRef] [PubMed]

35. Ackermann, D.; Schmitz, G. Detection and tracking of multiple microbubbles in ultrasound B-mode images. *IEEE Trans. Ultrason. Ferroelectr. Freq. Control* **2015**, *63*, 72–82. [CrossRef] [PubMed]
36. Christensenjeffries, K.; Browning, R.; Tang, M.X.; Dunsby, C.; Eckersley, R.J. In vivo acoustic super-resolution and super-resolved velocity mapping using microbubbles. *IEEE Trans. Med. Imaging* **2014**, *34*, 433–440. [CrossRef]

© 2018 by the authors. Licensee MDPI, Basel, Switzerland. This article is an open access article distributed under the terms and conditions of the Creative Commons Attribution (CC BY) license (http://creativecommons.org/licenses/by/4.0/).

Article

Ultrasonic Parametrization of Arterial Wall Movements in Low- and High-Risk CVD Subjects

Monika Makūnaitė [1],*, Rytis Jurkonis [1], Alberto Rodríguez-Martínez [2], Rūta Jurgaitienė [3], Vytenis Semaška [3], Karolina Mėlinytė [3] and Raimondas Kubilius [3]

1. Biomedical Engineering Institute, Kaunas University of Technology, 51423 Kaunas, Lithuania; rytis.jurkonis@ktu.lt
2. Communications Engineering Department, Miguel Hernandez University, 03202 Elche, Spain; arodriguezm@umh.es
3. Department of Cardiology, Hospital of Lithuanian University of Health Sciences (LSMU) Kauno klinikos, 50161 Kaunas, Lithuania; rutpuk@yahoo.com (R.J.); vytenis_semaska@yahoo.com (V.S.); karolina.melinyte@lsmuni.lt (K.M.); raimondas.kubilius@kaunoklinikos.lt (R.K.)
* Correspondence: makunaite.monika@gmail.com; Tel.: +370-602-45921

Received: 21 December 2018; Accepted: 26 January 2019; Published: 30 January 2019

Abstract: This paper shows the results of a preliminary study on the performance of new methods based on ultrasonic images parametrization, to estimate the arterial wall movements used for the evaluation of arterial stiffness, considered to be a predictor of cardiovascular events. The well-known technique of motion tracking in ultrasound image sequences was applied on cine loops scanned from subjects with different risks of suffering from cardiovascular disease (CVD). The motion of arterial walls was traced using displacement signals: Diameter, intima-media thickness (IMT) and longitudinal intima-media (IM) complex movement. The new methods used for the parametrization of the displacement signals were the average value (AV), effective or root mean square (RMS) value, and peak-to-peak motion amplitude estimate. A total of 79 subjects were analyzed in the study with 30 considered at low risk and 49 included in a preventive program for monitoring high CVD risk subjects. The results show a statistically significant difference between healthy volunteers and at-risk patients according to the AV of IMT, RMS values of longitudinal and radial motions and peak-to-peak amplitude of radial motion.

Keywords: common carotid artery; arterial wall motion; intima-media complex longitudinal motion; quantitative parametrization

---

1. Introduction

Cardiovascular diseases (CVDs) are the number one cause of human mortality and morbidity worldwide (WHO, 2017) [1]. Every year, more and more people die from these diseases than from any other illnesses. In 2016, 17.9 million people died from CVDs, constituting 31% of all global deaths. Heart attack and stroke make up 85% of these deaths [1] and the number of deaths from CVDs in the world is predicted to reach 23.6 million by 2030 [2].

CVDs are a group of disorders affecting the heart and blood vessels, which can cause myocardial infarction and stroke. They are usually acute events, mainly caused by a blood flow cut-off to the heart or brain, and described as the final stage of atherosclerosis [1,3], which is a systemic and chronic inflammatory disease of the medium and large arteries. Atherosclerosis is a degenerative progress that refers to the buildup over many years of lipids and other blood-borne materials in the arterial walls. Finally, atherosclerotic plaque forms, which can restrict blood flow in an artery. Overall, atherosclerotic arterial affection is not noticeable in the long term, but general signs of this disease develop only after complications start: Thickening of the intima-media complex, narrowing of the arterial lumen or its thrombosis, and/or loss of elasticity [4].

There are many risk factors that assist in the development of CVDs, which can be classified into two groups: Non-modifiable risk factors and modifiable risk factors. The first group's factors cannot be changed, and they are age, gender, family history, and race. The second group of factors can be changed or treated, and they include smoking, high blood pressure, diabetes, physical inactivity, overweight, high blood cholesterol, etc. [5]

An independent predictor of cardiovascular events is arterial stiffness. This parameter is generally analyzed to assess cardiovascular risk [5–7]. In clinical practice, the most commonly used risk markers for arterial stiffness evaluation are IMT, pulse wave velocity (PWV), and cross-sectional distensibility (CSD) [8–11]. Unfortunately, the clinical potential of these traditional risk markers as a screening test remains limited [12]. The risk of CVDs in patients under the age of 50 is difficult to evaluate, especially in the absence of specific individual CVDs risk factors or anamnesis. Therefore, it is difficult to assess the likelihood of developing a disease, and if so, to start drug treatment [5].

Nonetheless, it has been proven that significant anatomical changes (i.e., IMT) of the arterial wall appear much later than mechanical changes (i.e., longitudinal and radial motion of the arterial wall) [3]. Radial motion is a parameter describing the mechanical properties of the arterial walls, and it has been widely studied in recent years, becoming an informative non-invasive parameter that helps to investigate cardiovascular diseases and to determine the elasticity of arterial walls. Unlike the radial motion of the arterial wall, the longitudinal motion has not received such recognition. It was believed that the longitudinal motion during the heart cycle was negligible compared with the radial motion. However, using modern ultrasound scanners, it has been noticed that the innermost and middle layers of the large arteries (i.e., intima-media complex) during the heart cycle move not only in radial, but also in a longitudinal direction [13,14]. It has also been observed that the longitudinal motion of the arterial wall has the same amplitude as the radial motion and reaches about one millimeter [15]. In addition, clinical studies demonstrated the correlation of common carotid artery (CCA) longitudinal motion with risk factors and CVDs [16,17]. Previous studies have shown a relationship between the decrease in longitudinal motion of the CCA wall, arterial stiffness and CVDs [3]. While there is a link between longitudinal motion amplitude and CVDs, determinants of the phases of longitudinal motion remain unknown [9].

The aforementioned bidirectional longitudinal motion of the intima-media complex is observed during the heart cycle. Cinthio et al. [14] discuss the dependence of longitudinal motion peaks on heart cycle phases, i.e., systole and diastole. There are many speculations about what causes longitudinal motion in the arterial wall. Finally, determinants of the phases of longitudinal motion remains unknown [9]. At the beginning of systole, the first antegrade motion of the IM complex is observed, i.e., motion in the direction of blood flow. Later, still in systole, the first retrograde motion of this complex appears, i.e., motion in the opposite direction of blood flow. During diastole, the second antegrade motion of the IM complex follows and then it gradually returns to its original position [14]. Predominantly, only the longitudinal motion amplitude in different heart cycle phases is used. Most researchers measure the first antegrade, the first retrograde and the peak-to-peak amplitude of the longitudinal motion during the heart cycle [14,18,19]. Despite the fact that the longitudinal motion amplitude is used and is able to distinguish low risk (i.e., healthy controls) from high risk (i.e., at-risk patients), the entire longitudinal motion pattern (waveform) can be useful [19,20]. Moreover, the longitudinal motion pattern is different for different individuals [18,21]. To the best knowledge of the authors of this article, the RMS estimates for arterial wall movements were not tested in healthy controls and at-risk patients.

The aim of this paper is to evaluate both the motion average (AV) and RMS values for the parameterization of the arterial motion. Proposed parameters will be influenced by all amplitude values of the temporal variation of IMT, longitudinal and radial motion signals during the heart cycle.

2. Materials and Methods

2.1. Study Population

Thirty-three young healthy volunteers and sixty-nine older volunteers were involved in this study. Healthy subjects had no cardiovascular risk factors as assessed by a written questionnaire, while older subjects had high risk of cardiovascular diseases. The age of healthy volunteers was 22–23 years, while the at-risk patients' mean age was 51 years (±7 years standard deviation). In the control group, 9 subjects (27%) were male and 24 subjects (73%) were female. In the patients' group, 43 subjects (51%) were male and 26 subjects (49%) were female.

Clinical data were collected at the Lithuanian University of Health Sciences Hospital, Department of Cardiology, during May–October, 2018. The study was approved by the Kaunas Region Biomedical Research Ethics Committee (2018-08-02, No. BE-2-51, Kaunas, Lithuania). Every participant provided written consent to participate in the study and allowed the usage of the obtained B-mode images under the principle of confidentiality.

2.2. Collection of In Vivo Data

All analyses were performed using a clinical scanner Ultrasonix SonixTouch (Analogic Ultrasound, Canada), equipped with a 5–14 MHz linear array probe. During the CCA echoscopy, the frame rate, depth and focus were 52 fps, 2.5 cm and 2 cm, set in the ultrasound scanner accordingly. The data was stored in a cine-loop as consecutive frames for later offline analysis.

The acquisition of CCA B-mode sequences was performed by two cardiology physicians. Before the measurement, all subjects were asked to rest in supine position for at least 15 min. During the measurement, the subjects were lying in supine position, stretching their neck and turning it 45 degrees to the right or left, depending on the echoscopic neck side. Arterial longitudinal motion amplitude does not depend on the echoscopic neck side [8], so both right and left CCA were scanned. Measurements of the longitudinal movement and the diameter change of the CCA were performed 2–3 cm proximate to the bifurcation during at least two full heart cycles. In order to ensure that the CCA data was of acceptable quality, the longitudinal movement had to be clearly visible along the preselected segment of the arterial wall. All sequences were stored digitally and transferred to a computer for further analysis.

2.3. Estimation and Post-Processing of IMT, Longitudinal and Radial Motion Signals of CCA

For this study, CAROLAB software was used in order to estimate IMT, longitudinal and radial motions of CCA [10,22,23]. This software is used for the analysis of ultrasound B-mode image sequences and assesses the longitudinal motion with a speckle-tracking approach that is based on the block-matching (BM) method [10]. The main point of the BM framework is to detect the motion $d(n)$ between two consequent frames by comparing pixel blocks of consecutive images $I(n-1)$ and $I(n)$. The motion corresponds to the displacement between the center point $p(n-1)$ of the reference block and the center point $p(n)$ of the best-matched block. This results in the shift of the center point $p(n)$ between images $I(n-1)$ and $I(n)$. Pixel blocks alignment in images $I(n-1)$ and $I(n)$ takes place only within the search window, i.e., in the defined maximum margin around the center point of the reference and best-match blocks. After summing up all the displacements $d(n)$ received, the $p(n)$ point motion trajectory is estimated. In order to cope with the issue of speckle decorrelation, a pixel-wise Kalman filter is used to update the reference block [10]. Once the estimation is done, a fully-automatic technique based on front propagation is used [22] in order to segment the IM complex and track temporal variations in the IMT in CCA B-mode ultrasound images.

From right and left CCA image sequences, the higher-quality video sequence was chosen and finally one CCA image sequence was used for every subject. Subjects presenting low-quality video sequences or having less than two full heart cycles detected with CAROLAB were rejected from the study. Three subjects (9%) were rejected from the healthy volunteers' group while twenty subjects (29%) were rejected from the at-risk patients' group.

A region of interest (ROI) containing a well-contrasted speckle pattern of the distal vessel wall for longitudinal and radial motion, and clearly visible IM complex for IMT variation were chosen in the first frame of each B-mode sequence. A kernel of the ROI was selected manually, with size 3 × 0.5 mm as seen in Figure 1a. Estimated signals of longitudinal motion, radial motion and temporal variation in the IMT were saved for further post-processing in MATLAB.

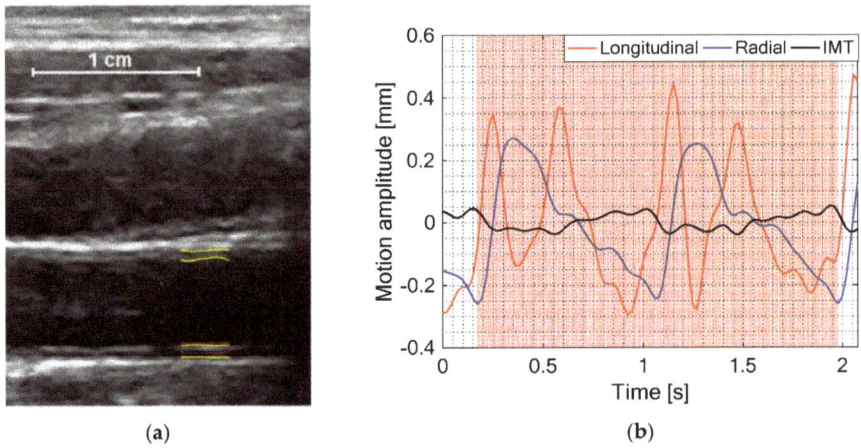

(a)   (b)

Figure 1. (**a**) Common carotid artery (CCA) echoscopy image with preselected kernel (red rectangular) for the estimation of the longitudinal motion and the segmented intima-media (IM) complex of both proximal and distal arterial walls (yellow lines) for temporal variation of intima-media thickness (IMT) and radial motion estimation in CAROLAB. (**b**) CAROLAB output signals detrended and filtered for better observation in a single diagram. A post-processed diameter signal was used to select two consequent heart cycles (red shadowed area). Only this time interval was used for quantitative parametrization of the signals.

Post-processing algorithm was developed in MATLAB. The CAROLAB output signals were loaded in original form as shown in Figure 2a.

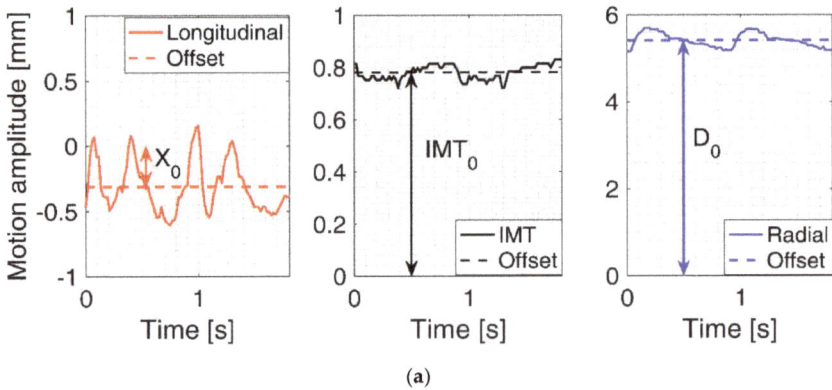

(**a**)

Figure 2. *Cont.*

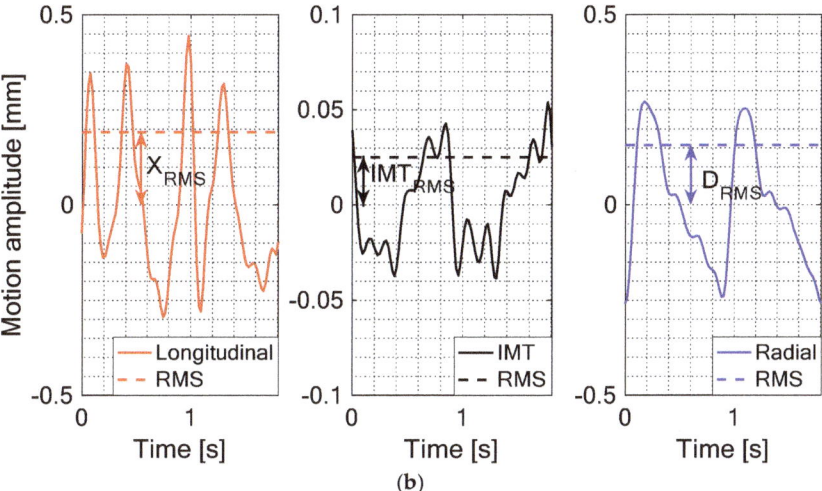

Figure 2. (a) Examples of three signals of CAROLAB output: Longitudinal motion, temporal variation of intima-media thickness (IMT) and radial motion. Average values (AV) or offsets of these three signals were denoted by X_0, IMT_0 and D_0 accordingly. (b) Post-processed (each motion signal was filtered and detrended, subtracting the offset of the signal) motion signals of the common carotid artery (CCA) wall were used for the estimation of the root mean square (RMS): X_{RMS}, IMT_{RMS}, D_{RMS}.

Motion signals were then filtered with a band-pass IIR filter ($f_{pass\text{-}lower}$ = 0.9 Hz and $f_{pass\text{-}higer}$ = 8 Hz) and then detrended, subtracting the mean of the resulting signal. After this, two consequent heart cycles were selected manually in time. Only this time segment (see red shadowed area in Figure 1b for an example) was used in all signals for the evaluation of time domain parameters: Peak-to-peak amplitude change of the arterial diameter and IMT between systole and diastole, AV and RMS values of temporal variation of IMT (IMT_0 and IMT_{RMS}), and longitudinal and radial motions (X_0, D_0 X_{RMS}, D_{RMS}) of two consequent heart cycles.

2.4. Quantitative Parametrization of IMT, Longitudinal and Radial Motion Signals of CCA

The average value (AV) is the mean amplitude of the waveform and can be calculated as follows:

$$AV = \frac{\sum_{i=1}^{n} x_i}{n} \quad (1)$$

The RMS value of a quantity is the square root of the mean value of the squared values of the quantity taken over an interval:

$$RMS = \sqrt{\frac{1}{n} \sum_{i=1}^{n} (x_i - DC)^2} \quad (2)$$

CAROLAB output signals were post-processed and quantitatively parametrized in our algorithm. From Equation (1), we calculate offset estimate or AV of longitudinal motion (X_0), diameter motion (D_0), and IMT motion (IMT_0) signals in the preselected time interval before post-processing. Examples of these estimates are shown with the help of arrows in Figure 2a. After filtering and detrending, all the aforementioned signals are parametrized calculating peak-to-peak amplitude and RMS estimates. Longitudinal motion RMS values were denoted by X_{RMS}, radial motions by D_{RMS}, and temporal variation of IMT by IMT_{RMS}. Examples of these estimates are indicated in Figure 2b.

2.5. Statistical Analysis

All the earlier calculated parameters were represented as box-and-whisker plots for evaluation and analysis. The data were assessed for normality using the Shapiro–Wilk test [24]. This test was applied to the two subject groups and for each evaluated parameter, separately. The significance level was 5% and we found that the data were not normally distributed. To determine the difference between the two groups, a Mann–Whitney U test was used. The value $p < 0.015$ was considered to indicate a statistically significant difference. Statistical analysis was performed using MATLAB.

3. Results

From now on, we will provide two sets of boxplots for comparisons: One for healthy volunteers (white) and one for at-risk patients (yellow).

Figure 3 shows the comparison between the AV of longitudinal motion, radial motion and temporal variation of IMT for healthy volunteers and at-risk patients. The AV of radial motion and IMT variation can be interpreted as arterial diameter and IMT at equilibrium instants between pulses. The AV of longitudinal motion is near zero because the IM complex is returning to the initial position after anterograde and retrograde longitudinal shifts. Both the radial and IMT average values in healthy volunteers are a little more closely grouped, whereas the variability of these measurements in the at-risk patients' group is slightly greater. Two extreme outliers of the IMT AV appear in both subjects' groups.

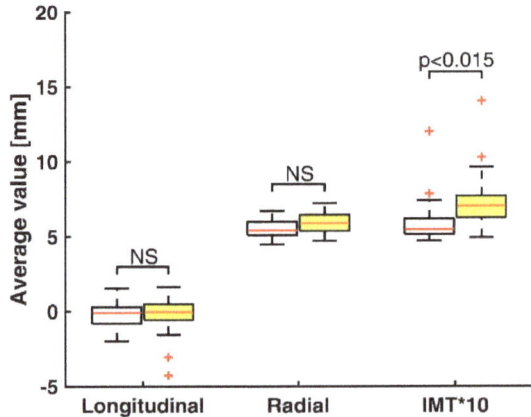

Figure 3. Box-and-whisker plots of average values (AV) of longitudinal motion (X_0), radial motion (D_0), and temporal variation of IMT (IMT_0) multiplied by a constant for healthy volunteers (white boxes, n = 30) and at-risk patients (yellow boxes, n = 49). Median values (successively, −0.09, −0.03, 5.41, 5.89, 5.49, 7.05) are shown as a horizontal red line within each box. Whiskers represent the minimum and maximum values. Outliers (estimates outside 1.5 times the inter-quartile range) are indicated by red +. The comparisons between the two groups are indicated by the p values. NS–non-significant.

Figure 4 shows statistics of RMS estimates of longitudinal motion, radial motion and temporal variation of IMT of the two groups. RMS estimates, or so-called effective values, are commonly used to indicate the time-averaged magnitude of a signal. In this particular case, this parameter accounts for the average activity of motion during two heart cycles. Both longitudinal and radial RMS values in healthy volunteers are a little wider whereas the spread of these measurements in at-risk patients are a little more closely grouped. Additionally, three to four extreme outliers of IMT RMS values appear in both groups.

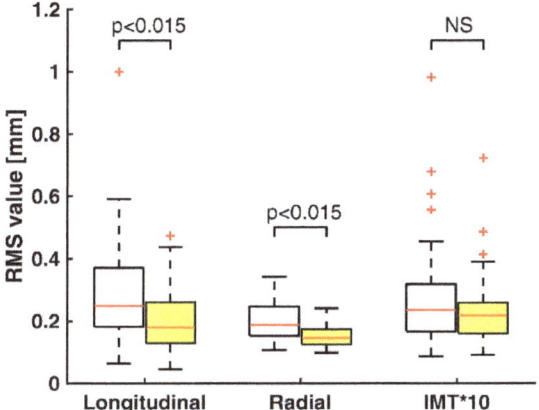

Figure 4. Box-and-whisker plots of root mean square (RMS) values of longitudinal motion (X_{RMS}), radial motion (D_{RMS}) and temporal variation of IMT (IMT_{RMS}) multiplied by a constant for healthy volunteers (white boxes, n = 30) and at-risk patients (yellow boxes, n = 49). Median values (successively, 0.25, 0.18, 0.19, 0.14, 0.24, 0.22) are shown as a horizontal red line within each box. Whiskers represent the minimum and maximum values. Outliers (estimates outside 1.5 times the inter-quartile range) are indicated by red +. The comparisons between the two groups are indicated by the *p* values. NS–non-significant.

Figure 5 shows the IMT and radial peak-to-peak motion amplitude distribution between the two groups. Peak-to-peak amplitudes represent time instant estimates of peaking motion signals. Radial motion signals are in maximal peak at systole and in minimal at diastole. IMT motion signals are in minimal peak at systole and in maximal at diastole. In addition, there are two extreme outliers in the healthy volunteers' group and three in the at-risk patients' group for the IMT peak-to-peak motion amplitude.

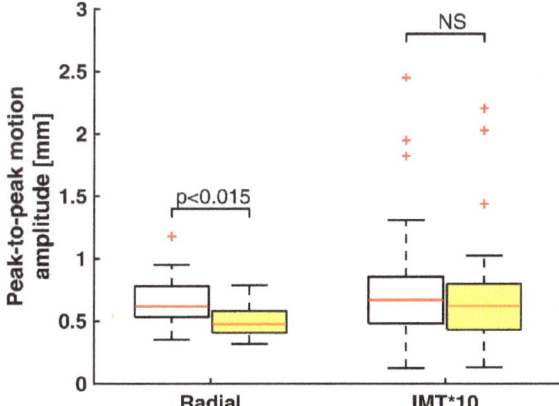

Figure 5. Box-and-whisker plots of peak-to-peak motion amplitude of radial motion and temporal variation of IMT multiplied by a constant for healthy volunteers (white boxes, n = 30) and at-risk patients (yellow boxes, n = 49). Median values (successively, 0.67, 0.62, 0.62, 0.47) are shown as a horizontal red line within each box. Whiskers represent the minimum and maximum values. Outliers (estimates outside 1.5 times the inter-quartile range) are indicated by red +. The comparisons between the two groups are indicated by the *p* values.

4. Discussion

In this study, we propose new parameters for the evaluation of the temporal variation of IMT and the longitudinal and radial arterial walls' motions during the heart cycle. The main purpose is to provide estimates that will be influenced by all amplitude values of the aforementioned motion variations.

We have used CAROLAB software [10,22,23] in order to analyze two different populations (33 healthy volunteers and 69 at-risk patients). Our results have demonstrated that the arterial diameter of healthy volunteers is not distinct from that of the at-risk patients, while there is a significant difference in the IMT between these two groups, as seen in Figure 3. The arterial diameter is the same among all subjects while IMT increases for at-risk patients. In accordance with previous studies [10,11], we can state that IM thickening is associated with CVDs. There is a significant difference between healthy volunteers and at-risk patients according to the RMS values of longitudinal and radial motions, as seen in Figure 4. From these results, it is clear that the healthy volunteers' CCA moves more in longitudinal and radial directions than in at-risk patients. This is consistent with what has been found in previous studies [17] and it may explain higher arterial elasticity in the healthy volunteers' group. In this paper we only have motion or displacement signals to analyze. Elastography researchers [25,26] state that in response to pulsatile flow, a stiffer artery moves less. This empirical knowledge about the negative correlation of motion amplitude with artery stiffness can be observed in our estimates. In addition, we have done our own experiments [27] determining that displacement correlates with stiffness. We found that with the decrease of agar-based phantoms' stiffness, the motion amplitude increases. No statistical difference can be claimed between the two groups according to the RMS value of IMT. Contrary to the findings of Zahnd et al. [23], we did not find any IMT variation increase in at-risk patients compared with healthy volunteers. Our results have demonstrated that there is no significant difference between healthy volunteers and at-risk patients according to IMT peak-to-peak motion amplitude, demonstrated in Figure 5. Although in our study two investigated populations were not so different, like healthy volunteers and diabetic patients, there is enough statistical difference of radial peak-to-peak motion amplitude between healthy volunteers and at-risk patients.

The main drawback is that some subjects presented low-quality B-mode sequences resulting in motion signals that were not repeatable. In these cases, the tracking process was repeated again with another kernel position in the CCA image, but this did not yield any better results. In addition, there were some sequences with fewer than two full heart cycles. Finally, all the B-mode sequences having such repeatability limitations were rejected from the study.

In order to get correct estimates of the temporal variation of IMT and radial motion, CCA walls have to appear as double-line patterns in all video sequences [14]. However, there were a few frames in some B-mode sequences where the intima-media complex was not as clearly visible as needed in CAROLAB software. This appears to be a case of errors resulting in inaccurate IM complex segmentation and finally incorrect values of IMT and radial motion.

According to our study, the echoscopy of CCA for the registration of temporal variation of IMT, longitudinal and radial motions is challenging. Following completion of the work reported in this paper, Au et al. [28] suggested to average four consequent heart cycles for representative measurement of longitudinal motion in CCA. Au et al. [28] noticed that indices of variability were reduced when two to four heart cycles were used. Any improvements in indices of variability were not observed when more than four heart cycles were averaged. This proves that more than four heart cycles are unnecessary to use [28]. In future studies, we propose using four heart cycles for representative estimates of not only longitudinal, but also temporal variation of IMT and radial motion. In addition, parameters for consecutive heart cycles' repeatability must be incorporated into the B-mode sequences' acquisition. Only similar consecutive heart cycles will be taken for evaluation of CCA motion. Standard deviation values can be derived between four consequent heart cycles as feedback for the sonographer and determine if the acquired sequence is acceptable or not. As we understand it, this gives clearly better results and ensures correct parametrization of the temporal variation of IMT, longitudinal and radial motions.

AV and RMS values of IMT variation, longitudinal and radial motions of two consequent heart cycles are new parameters, which have not been used previously in clinical studies associated with CCA and atherosclerosis. Future investigations are necessary to validate the conclusions that can be drawn from this study. Frequency domain parameters could be proposed in the future to distinguish healthy subjects from at-risk patients.

5. Conclusions

Reliable recording of sequences of echoscopy images from subject CCA is challenging. Collaboration of the subject is necessary to keep the body motionless during recording. All subjects' motions during echoscopy were recorded as artefacts and this decreased repeatability of arterial pulsing movements. The results show statistically significant difference between healthy volunteers and at-risk patients according to the AV of temporal variation of IMT, RMS values of longitudinal and radial motions, and peak-to-peak amplitude of radial motions.

Author Contributions: Conceptualization, R.J. (Rytis Jurkonis) and R.K.; data curation, V.S.; formal analysis, M.M.; funding acquisition, R.J. (Rytis Jurkonis) and R.K.; investigation, R.J. (Rūta Jurgaitienė) and V.S.; methodology, R.J. (Rytis Jurkonis); project administration, R.J. (Rytis Jurkonis) and K.M.; resources, R.J. (Rytis Jurkonis) and K.M.; software, M.M. and R.J. (Rytis Jurkonis); supervision, R.J. (Rytis Jurkonis) and R.K.; visualization, M.M.; writing—original draft, M.M. and R.J. (Rytis Jurkonis); writing—review and editing, A.R.-M.

Funding: This research was supported by the Research, Development and Innovation Fund of Kaunas University of Technology (grant no. PP22/185) and the Research Fund of Lithuanian University of Health Sciences (grant no. BN18-73/SV5-0296).

Conflicts of Interest: The authors declare no conflict of interest.

References

1. Cardiovascular Diseases (CVDs). Available online: https://www.who.int/en/news-room/fact-sheets/detail/cardiovascular-diseases-(cvds) (accessed on 11 December 2018).
2. Cardiovascular Diseases (CVDs)–Global Facts and Figures. Available online: https://www.world-heart-federation.org/resources/cardiovascular-diseases-cvds-global-facts-figures/ (accessed on 17 December 2018).
3. Zahnd, G.; Boussel, L.; Sérusclat, A.; Vray, D. Intramural Shear Strain can Highlight the Presence of Atherosclerosis: A Clinical in Vivo Study. In Proceedings of the 2011 IEEE International Ultrasonics Symposium, Orlando, FL, USA, 18–21 October 2011; pp. 1770–1773.
4. Molinari, F.; Suri, J.S. Automated Measurement of Carotid Artery Intima-Media Thickness. In *Ultrasound and Carotid Bifurcation Atherosclerosis*; Nicolaides, A., Beach, K.W., Kyriacou, E., Pattichis, C.S., Eds.; Springer: London, UK; Dordrecht, The Netherlands; Heidelberg, Germany; New York, NY, USA, 2012; pp. 177–192, ISBN 978-1-84882-687-8.
5. Hobbs, F.; Hoes, A.W.; Agewall, S.; Albus, C.; Brotons, C.; Catapano, A.L.; Cooney, M.; Corrà, U.; Cosyns, B.; Deaton, C.; et al. 2016 European Guidelines on Cardiovascular Disease Prevention in Clinical Practice. *Eur. Heart J.* **2016**, *37*, 2315–2381. [CrossRef]
6. Zahnd, G.; Orkisz, M.; Sérusclat, A.; Vray, D. Minimal-Path Contours Combined with Speckle Tracking to Estimate 2D Displacements of the Carotid Artery Wall in B-Mode Imaging. In Proceedings of the 2011 IEEE International Ultrasonics Symposium, Orlando, FL, USA, 18–21 October 2011; pp. 732–735.
7. Ahlgren, A.R.; Cinthio, M.; Steen, S.; Persson, H.W.; Sjöberg, T.; Lindström, K. Effects of Adrenaline on Longitudinal Arterial Wall Movements and Resulting Intramural Shear Strain: A First Report. *Clin. Physiol. Funct. Imaging* **2009**, *29*, 353–359. [CrossRef] [PubMed]
8. Svedlund, S.; Gan, L. Longitudinal Wall Motion of the Common Carotid Artery can be Assessed by Velocity Vector Imaging. *Clin. Physiol. Funct. Imaging* **2011**, *31*, 32–38. [CrossRef] [PubMed]

9. Au, J.S.; Ditor, D.S.; Macdonald, M.J.; Stöhr, E.J. Carotid Artery Longitudinal Wall Motion is Associated with Local Blood Velocity and Left Ventricular Rotational, but Not Longitudinal, Mechanics. *Physiol. Rep.* **2016**, *4*, 1–11. [CrossRef] [PubMed]
10. Zahnd, G.; Orkisz, M.; Sérusclat, A.; Moulin, P.; Vray, D. Evaluation of a Kalman-Based Block Matching Method to Assess the Bi-Dimensional Motion of the Carotid Artery Wall in B-Mode Ultrasound Sequences. *Med. Image Anal.* **2013**, *17*, 573–585. [CrossRef] [PubMed]
11. Bauer, M.; Caviezel, S.; Teynor, A.; Erbel, R.; Mahabadi, A.A.; Schmidt-Trucksäss, A. Carotid Intima-Media Thickness as a Biomarker of Subclinical Atherosclerosis. *Swiss Med. Wkly.* **2012**, *142*, 1–9. [CrossRef] [PubMed]
12. Zahnd, G.; Salles, S.; Liebgott, H.; Vray, D.; Sérusclat, A.; Moulin, P. Real-time Ultrasound-tagging to Track the 2D Motion of the Common Carotid Artery Wall in Vivo. *Med. Phys.* **2015**, *42*, 820–830. [CrossRef] [PubMed]
13. Persson, M.; Ahlgren, A.R.; Jansson, T.; Eriksson, A.; Persson, H.W.; Lindström, K. A New Non-invasive Ultrasonic Method for Simultaneous Measurements of Longitudinal and Radial Arterial Wall Movements: First in Vivo Trial. *Wiley Online Libr.* **2003**, *23*, 247–251. [CrossRef]
14. Cinthio, M.; Ahlgren, A.R.; Bergkvist, J.; Jansson, T.; Persson, H.W.; Lindström, K. Longitudinal Movements and Resulting Shear Strain of the Arterial Wall. *Am. J. Physiol.-Heart Circ. Physiol.* **2006**, *291*, 394–402. [CrossRef] [PubMed]
15. Cinthio, M.; Ahlgren, A.R.; Jansson, T.; Eriksson, A.; Persson, H.W.; Lindstrom, K. Evaluation of an Ultrasonic Echo-Tracking Method for Measurements of Arterial Wall Movements in Two Dimensions. *IEEE Trans. Ultrason. Ferroelectr. Freq. Control* **2005**, *52*, 1300–1311. [CrossRef] [PubMed]
16. Svedlund, S. Carotid Artery Longitudinal Displacement Predicts 1-Year Cardiovascular Outcome in Patients with Suspected Coronary Artery Disease Arteriosclerosis. *Thromb. Vasc. Biol.* **2011**, *31*, 1668–1674. [CrossRef] [PubMed]
17. Zahnd, G.; Boussel, L.; Marion, A.; Durand, M.; Moulin, P.; Sérusclat, A.; Vray, D. Measurement of Two-Dimensional Movement Parameters of the Carotid Artery Wall for Early Detection of Arteriosclerosis: A Preliminary Clinical Study. *Ultrasound Med. Biol.* **2011**, *37*, 1421–1429. [CrossRef] [PubMed]
18. Taivainen, S.H.; Yli-Ollila, H.; Juonala, M.; Kähönen, M.; Raitakari, O.T.; Laitinen, T.M.; Laitinen, T.P. Interrelationships between Indices of Longitudinal Movement of the Common Carotid Artery Wall and the Conventional Measures of Subclinical Arteriosclerosis. *Clin. Physiol. Funct. Imaging* **2017**, *37*, 305–313. [CrossRef] [PubMed]
19. Yli-Ollila, H.; Laitinen, T.; Weckström, M.; Laitinen, T.M. New Indices of Arterial Stiffness Measured from Longitudinal Motion of Common Carotid Artery in Relation to Reference Methods, a Pilot Study. *Clin. Physiol. Funct. Imaging* **2016**, *36*, 376–388. [CrossRef] [PubMed]
20. Qorchi, S. Extraction of Characteristic Patterns from Carotid Longitudinal Kinetics. In Proceedings of the Recherche en Imagerie et Technologies pour la Santé (RITS) 2017, Lyon, France, 27–29 March 2017.
21. Cinthio, M.; Albinsson, J.; Erlöv, T.; Bjarnegård, N.; Länne, T.; Ahlgren, Å.R. Longitudinal Movement of the Common Carotid Artery Wall: New Information on Cardiovascular Aging. *Ultrasound Med. Biol.* **2018**, *44*, 2283–2295. [CrossRef] [PubMed]
22. Zahnd, G.; Orkisz, M.; Vray, D. *Imaging-Based Computational Biomedicine Lab*; Graduate School of Information Science, Nara Institute of Science and Technology (NAIST): Nara, Japan, 2017. [CrossRef]
23. Zahnd, G.; Kapellas, K.; van Hattem, M.; van Dijk, A.; Sérusclat, A.; Moulin, P.; van der Lugt, A.; Skilton, M.; Orkisz, M. A Fully-Automatic Method to Segment the Carotid Artery Layers in Ultrasound Imaging: Application to Quantify the Compression-Decompression Pattern of the Intima-Media Complex during the Cardiac Cycle. *Ultrasound Med. Biol.* **2017**, *43*, 239–257. [CrossRef] [PubMed]
24. Razali, N.M.; Wah, Y.B. Power comparisons of Shapiro-Wilk, Kolmogorov-Smirnov, Lilliefors and Anderson-Darling tests. *J. Stat. Model. Anal.* **2011**, *2*, 21–33.
25. Wells, P.N.T.; Liang, H.-D. Medical ultrasound: Imaging of soft tissue strain and elasticity. *J. R. Soc. Interface* **2011**, *8*, 1521–1549. [CrossRef] [PubMed]
26. Shiina, T.; Nightingale, K.R.; Palmeri, M.L.; Hall, T.J.; Bamber, J.C.; Barr, R.G.; Castera, L.; Choi, B.I.; Chou, Y.H.; Cosgrove, D.; et al. WFUMB guidelines and recommendations for clinical use of ultrasound elastography: Part 1: Basic principles and terminology. **2015**, *41*, 1126–1147. [CrossRef]

27. Zambacevičienė, M. RF Ultrasound Based Estimation of Pulsatile Flow Induced Microdisplacements in Phantom. In Proceedings of the World Congress on Medical Physics and Biomedical Engineering 2018, Prague, Czech Republic, 3–8 June 2018; pp. 601–605.
28. Au, J.S.; Yli-Ollila, H.; Macdonald, M.J. An Assessment of Intra-Individual Variability in Carotid Artery Longitudinal Wall Motion: Recommendations for Data Acquisition. *Physiol. Meas.* **2018**, *39*, 09NT01. [CrossRef] [PubMed]

 © 2019 by the authors. Licensee MDPI, Basel, Switzerland. This article is an open access article distributed under the terms and conditions of the Creative Commons Attribution (CC BY) license (http://creativecommons.org/licenses/by/4.0/).

MDPI
St. Alban-Anlage 66
4052 Basel
Switzerland
Tel. +41 61 683 77 34
Fax +41 61 302 89 18
www.mdpi.com

Applied Sciences Editorial Office
E-mail: applsci@mdpi.com
www.mdpi.com/journal/applsci

www.ingramcontent.com/pod-product-compliance
Lightning Source LLC
LaVergne TN
LVHW071956080526
838202LV00064B/6761